Making
Printed Circuit
Boards

Making
Printed Circuit
Boards

Jan Axelson

TAB Books
Imprint of McGraw-Hill

New York San Francisco Washington, D.C. Auckland Bogotá
Caracas Lisbon London Madrid Mexico City Milan
Montreal New Delhi San Juan Singapore
Sydney Tokyo Toronto

© 1993 by **Janet Louise Axelson**.
Published by TAB Books.
TAB Books is a division of McGraw-Hill, Inc.

pbk   12 13 14 FGR/FGR 0 5 4 3
hc   1 2 3 4 5 6 7 8 9 10 FGR/FGR 9 9 8 7 6 5 4 3

Library of Congress Cataloging-in-Publication Data

Axelson, Janet Louise.
    Making printed circuit boards / by Janet Louise Axelson.
        p.   cm.
    Includes bibliographical references and index.
    ISBN 0-8306-3951-9 (pbk.) ISBN 0-8306-3950-0 (hard)
    1. Printed circuits—Design and construction. I. Title.
TK7868.P7A97 1993
621.3815'31—dc20   93-9149
                                            CIP

Acquisitions editor: Roland S. Phelps
Editorial team: Laura J. Bader, Editor
                       Susan Wahlman, Managing Editor
                       Joanne Slike, Executive Editor
                       Joann Woy, Indexer
Design team: Jaclyn J. Boone, Designer
                       Brian Allison, Associate Designer
Cover design: Holberg Design, York, Pa.   EL2
Cover photograph: Thompson Photography, Baltimore, Md.   4142

to Jim

Contents

Acknowledgments

This book was made possible by the assistance of dozens of companies that generously provided information, photos, and product samples for my research. Special thanks to Capital Electro-Circuits, Dremel, DynaArt Designs, ESP Solder Plus, Excellon Automation Division of Esterline Corporation, General Consulting, Gerber Scientific Inc., PACE Inc., PADS Software, Planned Products, Solder World, Techniks Inc., and Ungar. Thanks also to Jim for meeting every photographic challenge.

Trademarks

AutoCAD is a registered trademark and DXF is a trademark of Autodesk, Incorporated

Circuit Works is a trademark of Planned Products

CIR-KIT and PACE are registered trademarks of PACE, Incorporated

Datak, Pos-Neg, and TINNIT are registered trademarks of The Datak Corporation

Dot-Maker, ESP, and VAC Tweezer are registered trademarks of ESP Solder Plus

Ersin is a registered trademark of Multicore Solders

Excellon is a registered trademark of Excellon Automation

Generic CADD is a registered trademark of Autodesk Retail Products

Gerber is a registered trademark of Gerber Scientific Incorporated

HP-GL is a registered trademark of Hewlett-Packard Company

IBM, IBM PC, and OS/2 are registered trademarks of International Business Machines, Incorporated

Kodalith, D-11, and Kodak are trademarks of Eastman Kodak Company

Macintosh is a trademark of Apple Computer

Microsoft Windows and MS-DOS are registered trademarks of Microsoft Corporation

Mylar is a registered trademark of DuPont Company

OrCAD is a registered trademark of OrCAD Systems Corporation

PADS is a registered trademark of PADS Software

PCS is a registered trademark of GC Thorsen

Postscript is a registered trademark of Adobe Systems, Incorporated
Press-n-Peel is a registered trademark of Techniks Incorporated
Scotch is a trademark and Scotchflex is a registered trademark of 3M Corporation
TEC-200 is a trademark of the Meadowlake Corporation
ulano is a registered trademark of The Ulano Companies

Introduction

Electronic circuits built on printed circuit (pc) boards have several advantages: they're compact, professional-looking, easy to troubleshoot and repair, and easy to duplicate. This is why just about everyone who builds or experiments with electronic circuits eventually becomes interested in making pc boards.

The good news is that making your own pc boards isn't hard to do. You can get top-quality results using a few basic tools and materials, and move on to more complex techniques when you need to—for example, when circuit complexity increases or when you need to make your boards in larger quantities.

Who should read this book?

To guide you in making your own pc boards, this book takes you step by step from schematic diagram to soldered, ready-to-use board. If you've never made a pc board before, you'll find the information here to get you started making your own boards right away. If you already have experience in making pc boards, you'll find new ideas on how to design and fabricate boards, how to improve their quality, and how to make them more efficiently.

Whether you're an electronics hobbyist, technician, circuit designer, prototype builder, engineer, student, or just someone who is interested in or involved with electronic circuits or pc boards, the information in this book will help you in designing, making, repairing, and using pc boards.

This book came about because of my own curiosity about and interest in finding an easy, quick, effective, and low-cost method for making pc boards for my own projects. In my research, I contacted dozens of vendors of pc board-related products, and I experimented with all kinds of processes and procedures. This book presents the results of my research.

How to use this book

This book will guide you in each of these steps in pc board fabrication:

- Drawing a complete and accurate schematic diagram.
- Creating the pc board artwork.
- Transferring the artwork to the board.
- Etching the copper pattern onto the board.
- Drilling holes for component leads and mounting hardware.
- Soldering components to the board.
- Repairing and modifying pc boards.
- Building three sample projects.

You can read the book straight through or skip to whatever topics interest you. Step-by-step procedures show exactly what's involved in performing many of the processes described. For many tasks, you can choose from a variety of methods, depending on whether your goal is low cost, simplicity, quick results, convenience, or another consideration.

This book is also intended as a reference that you can refer to as needed, for example, when you are ready to make your first double-sided board, or when you are looking for a more efficient or effective method of image transfer.

The book's focus is on small-scale fabrication—for when you need anywhere from one to several dozen copies of a board of one or two etched layers.

Some of the questions answered in this book include

- What kinds of circuits can I build on a pc board?
- How can I ensure that my schematic diagrams include all of the information I'll need in building and testing my circuits?
- What is a simple, quick, and low-cost way to make good-quality pc boards?
- Can I use a personal computer to help draw the schematic diagram and pc board artwork, without spending a fortune on software?

- How do I design and make a board using the new surface-mount components?
- How do I transfer my pc board artwork from my computer to the pc board?
- What is an easy way to make a board using the pc board artwork published in books and magazines?
- How can I make a double-sided pc board with accurate registration between the top and bottom layers?
- Where can I buy the equipment and materials I need?
- What are some simple and effective techniques for pc board repair?

What's inside

Chapter by chapter, here's what you'll find inside this book.

Chapter 1 introduces you to pc boards, including what they are, when to use them (and when not to), and their alternatives. An overview of the pc board fabrication process shows what's involved in making a pc board.

Chapter 2 presents techniques for drawing complete and accurate schematic diagrams that will help you build, test, and use your circuits. If you have a personal computer, you'll find out how you can use it to draw schematics using dedicated schematic capture software as well as general-purpose CAD (computer-aided design) software. If you are in the market for schematic drawing software, the included checklists will help you find the best product for your needs.

Chapter 3 covers component choices that relate to pc board design, including how to decide whether to use the new surface-mount or traditional through-hole components.

Chapter 4 presents guidelines for designing the artwork that will be etched in copper on your pc board. Topics covered include how to select trace widths and pad sizes, how to route traces successfully, techniques for designing double-sided boards, and what to include in a parts placement diagram and other supplementary artwork.

Chapter 5 illustrates a variety of ways to create the pc board artwork, including traditional transfer tapes and patterns, and computer-aided design using dedicated pc board design software and general-purpose CAD software. A buyer's guide helps you choose pc board design software to suit your needs and budget.

Chapter 6 includes methods for converting artwork from one format to another in preparation for image transfer, how to choose pc board base materials, and how to cut and clean a blank pc board.

Chapter 7 describes methods for transferring the artwork to the pc board, including laying tapes and pads, iron-on image transfer, plotting directly onto copper, and contact printing using light-sensitive materials. Also included are techniques for making double-sided boards.

Chapter 8 describes how to etch the transferred image onto the pc board, including etchant choices and how to etch effectively and safely.

Chapter 9 is a guide to drilling and soldering. Topics covered include what kind of drill and drill bits to use, drilling techniques, and how to solder reliable connections to both through-hole and surface-mount components.

Chapter 10 illustrates how to make neat and functional repairs to your boards.

Chapter 11 presents three complete projects that illustrate pc board design for surface-mount and through-hole components. You can use these projects as a starting point for practicing the techniques described in this book, and then move on to designing pc boards for your own designs.

Finally, when you're ready to buy materials or supplies, the appendices contain comprehensive lists of manufacturers and vendors of software, materials, supplies, and equipment relating to pc boards. Also included are lists of books and magazines relating to pc boards, which you can consult to learn more or to keep up on the latest developments.

I hope you find this book to be a useful and helpful guide and reference in your pc board experiments and projects. If you have comments, corrections, or suggestions that might be useful for future editions of this book, please send them to me c/o TAB Books, Blue Ridge Summit, PA 17294-0850.

1
Printed circuit board basics

This chapter introduces you to printed circuit (pc) boards and includes basic definitions, an overview of how pc boards are made, their advantages and disadvantages, and how to decide whether to use a pc board or a different construction method for a particular project. Also included is important information on safeguarding your health, safety, and environment when you are fabricating or working with pc boards.

Basic definitions

Printed circuit boards are used in all kinds of electronic circuits, from simple one-transistor amplifiers to the biggest super-computers. You can find pc boards in cars, telephones, ovens, toys, televisions, computers, and much more. Figure 1-1 shows a variety of commercially fabricated pc boards with components installed.

This book concentrates on how to design and make single-sided and double-sided rigid pc boards. These boards are versa-tile enough for all but the most complex or critical designs, and you can make them without a large investment in special tools or materials. A single-sided board consists of a flat panel of insu-lating base material with a pattern drawn in copper foil on one side, while a double-sided board has patterns on both sides.

The copper pattern on a pc board is made up primarily of two shapes: *pads,* also called *lands,* which connect to *traces,* or *tracks,* as Fig. 1-2 illustrates. The pads can be round, rectangular, oval, or other shapes and provide a surface for soldering com-ponent leads or terminations. Some pads have small holes in

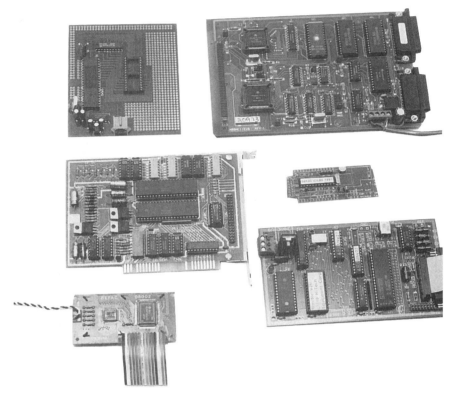

Fig. 1-1 *Pc boards vary widely in size, shape, and complexity, as shown by this assortment of commercially fabricated boards.*

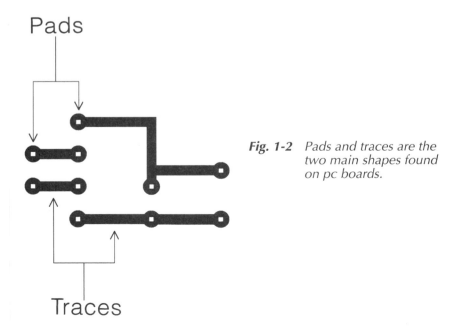

Pads

Fig. 1-2 *Pads and traces are the two main shapes found on pc boards.*

Traces

their centers for component leads. Other pads have no lead holes; the component leads or terminations rest on top of the pad. The traces are the conductors that interconnect the pads. A set of interconnecting pads and traces is often referred to as the pc board *artwork.*

Components soldered to a pc board might include integrated circuits (ICs), resistors, capacitors, inductors, connectors, or just about any element that an electronic circuit might contain. When the components are soldered to their pads, the traces connect the leads and create a working circuit. The base material acts as a support and backing material for the copper pattern. It also acts as an electrical insulator to isolate individual traces and pads from each other.

The name *printed circuit board* suggests that printing processes are involved in drawing the artwork on the board. And printing processes are often used to transfer an image to a pc board. But the actual copper image is usually created not by printing, but by chemically etching away unwanted copper from a copper-coated surface. Other terms that are sometimes used to describe pc boards include *printed wiring boards* and *etched wiring boards.*

Types of boards

A pc board can be single-sided, double-sided, or multilayer. Figure 1-3 illustrates the differences. A single-sided board has artwork on one side only, while a double-sided board has artwork on both sides. On a double-sided board, connections between layers are made with small, conductive holes called *vias* that connect to a pad on each side of the board.

A multilayer board has artwork on one or more internal layers in addition to the top and bottom. You can think of a multilayer board as a stack of pc boards, with a double-sided board on the bottom and one or more single-sided boards laminated on top.

Multilayer boards are difficult to fabricate without special equipment to precisely align the layers. If you need a multilayer board, you can design the artwork using the techniques described in this book, and send the board out to be fabricated by a shop that has the necessary equipment and materials.

History of pc boards

The technology for making pc boards was invented in the 1930s and came into use during World War II. Before that time, circuits were constructed with point-to-point soldering—

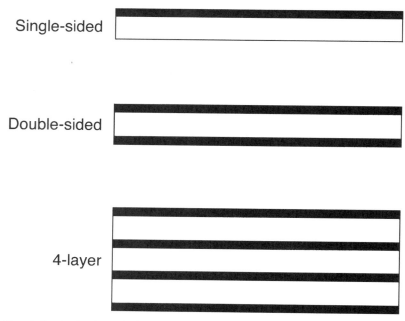

Single-sided

Double-sided

4-layer

Fig. 1-3 *A single-sided pc board has artwork drawn in copper on just one side of the board, while a double-sided board has copper layers on both sides, and a multilayer board has several layers, each separated by insulating base material.*

components are mounted on an insulating board and interconnecting wires are hand-soldered to the component leads.

Point-to-point soldering is a feasible method of circuit construction, but it's time-consuming and prone to wiring errors. Large hand-wired circuits are also bulky and hard to troubleshoot because of their many wires. In contrast, circuits built on pc boards are neat and easy to solder, duplicate, and troubleshoot.

Six steps to pc board fabrication

Before getting into the specifics of how to design and make a pc board, let's first look at the overall process to get an idea of how each step contributes to the final board.

Most traditional methods for making pc boards use subtractive techniques, where excess copper is removed from a copper-coated board, leaving the artwork drawn on an otherwise bare board. Using typical subtractive methods, there are six main steps in making a pc board:

1. Draw or obtain a schematic diagram of the circuit.
2. Design the pc board artwork.
3. Transfer an image of the artwork to the circuit board.
4. Etch the artwork in copper on the board.
5. Drill holes for component leads, connectors, and mounting hardware.
6. Solder components to the board.

You're then ready to power up, test, and begin using the circuit. Figure 1-4 shows a board during all six stages of design and fabrication.

Now let's look at each step in greater detail.

The schematic

The schematic diagram is a picture that describes a circuit's components and how they interconnect. The schematic diagram for a circuit can be obtained in a variety of ways. You can design the circuit and draw the schematic yourself, or you can use an existing schematic from a book, magazine, or other source. If you have a computer, you can draw the schematic with the aid of computer-aided design (CAD) software.

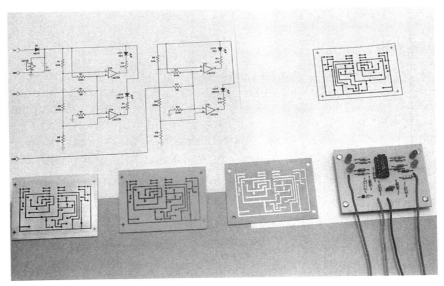

Fig. 1-4 *The six stages of pc board fabrication. The schematic and pc board artwork (top) and boards after image transfer, etching, drilling, and soldering (bottom).*

Whatever the source of your schematic, before you begin the artwork, you should check the schematic carefully and verify that its words and symbols contain a complete and accurate description of the circuit you want to build. Often it's a good idea to build and test a prototype of your circuit using a solderless breadboard or other method before you commit yourself to a pc board layout. A working prototype helps to verify that your design is sound and your schematic is accurate.

The artwork

The pc board artwork, or layout, is the pattern that is etched in copper on your pc board. For many years, drafters have created artwork using special opaque press-on or rub-on patterns and tapes to lay out artwork on drafting film. These methods are still being used, but as with schematic drawing, computer-aided design is replacing manual methods of pc board design.

With computer-aided design, instead of laying out the artwork on drafting film, you first draw the artwork on a computer screen. When the artwork is complete, you can print or plot it on paper or transparent film, or you can generate a computer file in a special format used by a photoplotter to draw a high-quality image of the artwork on a light-sensitive medium.

If you are using artwork from a book, magazine, or other existing artwork, you don't have to worry about designing the artwork—this step is already done for you.

Image transfer

When the artwork is complete, the next step is to transfer the image onto a copper-coated pc board. The transferred image must be etch resistant. This means that it cannot react chemically with the etching solution, which removes all exposed copper, leaving the artwork protected by the etch resist on an otherwise bare board. Figure 1-5 shows an unetched board with artwork transferred to it.

There are several methods for transferring an image to a pc board. One simple method is to apply rub-on or press-on patterns directly onto a copper-coated board. Another option is to photocopy or laser print the artwork onto a special transfer sheet and use heat and pressure to transfer the image to the board. You can also use photographic techniques to expose the artwork onto a board coated with a photosensitive material and develop the image. You can even plot the artwork in resist ink directly onto a copper sheet that you then laminate to a base material. Screen

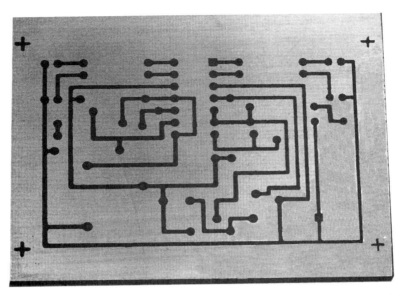

Fig. 1-5 *A pc board with artwork transferred onto it, ready for etching.*

printing offers yet another option for quick image transfer to multiple boards. Each of these methods creates an etch-resistant pattern on the board.

Etching

In the etching step, the board is immersed in a chemical bath that removes all the exposed copper. When etching is complete, the board is bare except for the circuit pattern drawn in copper, as Fig. 1-6 shows.

Drilling

After etching, the board is ready for cleaning and drilling of component and mounting holes, as shown in Fig. 1-7.

Soldering

After the holes have been drilled, you're ready to insert or mount the components on the pc board and solder them to their pads, as shown in Fig. 1-8. The board is then ready for testing, installing in an enclosure, and use.

Putting it all together

As you can see, designing and making pc boards requires you to be a master of many trades. You use technologies and processes

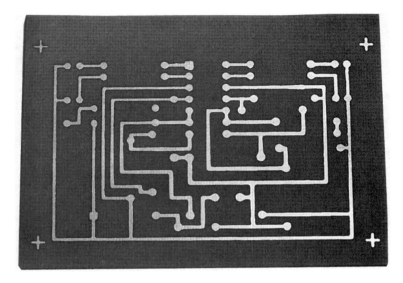

Fig. 1-6 *An etched pc board.*

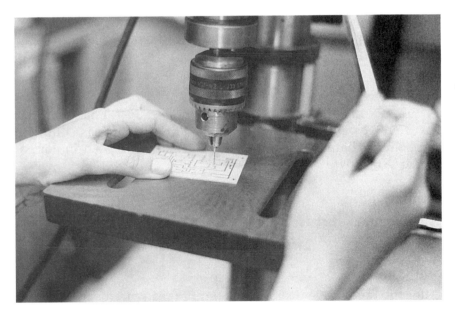

Fig. 1-7 *Drilling lead holes in a pc board.*

drawn from electronics, drafting, chemistry, photography, print-ing, and machine shop, to name a few.

Many circuit designers enjoy making their own boards from start to finish. Doing it all yourself gives you complete control,

Fig. 1-8 *Soldering components to a pc board.*

and can also save you time and money. But you don't have to do everything. For example, you might make your own board from existing artwork, or you might design the artwork for a complex board, but hire out the board fabrication. Any steps that you don't care to do, or aren't equipped to do, you can delegate to specialists.

Alternatives to pc boards

A pc board isn't the only way to construct a circuit. Alternate circuit fabrication methods include solderless breadboarding, wire wrapping, and point-to-point soldering.

For mass production, pc boards are the only practical method of circuit construction, and they are often desirable even for a single device or just a few copies of a circuit. But in some situations, you might find another method of circuit construction to be easier, quicker, cheaper, or more convenient than making a pc board. In addition, alternate methods are useful for prototyping a design before you commit yourself to making a pc board.

The following sections take a brief look at pc board alternatives.

Solderless breadboarding

When you want to try out a circuit idea, or build a circuit that you will need for only a few days or weeks, a solderless bread-

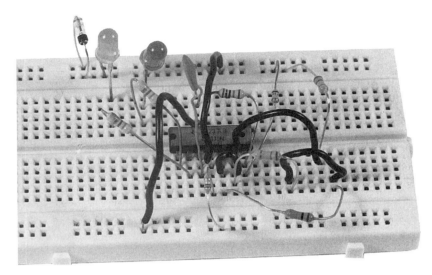

Fig. 1-9 *Simple circuits can be constructed for temporary use on a solderless breadboard.*

board is often a good solution. Figure 1-9 shows a circuit constructed on a solderless breadboard. The breadboard contains rows of sockets into which you plug component legs and jumper wires. Within a row, all the sockets connect electrically, so component leads can be connected just by inserting them into connecting sockets.

You can easily swap components and change the wiring by inserting and removing wires and components. When you're finished with the circuit, you can remove the components and reuse them and the breadboard in new circuits.

Solderless breadboards are best for building temporary versions of simple circuits. The plug-in connections are great for experimenting, but unsuitable for long-term use, because dropping or jarring the board can cause components or wires to come loose or short together. Larger, more complex circuits are unwieldy on a breadboard and require more permanent, durable connections. In addition, the long lead lengths make solderless breadboards unsuitable for critical high-frequency designs, and breadboard sockets do not accept surface-mount components without special adapters.

Wire wrapping

Wire wrapping is a more durable construction method than solderless breadboarding, yet it also allows for modifications.

Fig. 1-10   *A circuit constructed using wire-wrap tools and materials.*

Figure 1-10 shows a circuit put together with wire-wrap materials and techniques.

To wire wrap a circuit, you need special wire-wrap sockets with long, square pins, or posts. The sockets are inserted into perforated board, but instead of soldering the connections, you use a special tool to wrap several turns of wire around the socket pins. The stripped end of the wire forms a durable electrical and mechanical connection to the pin.

A simple hand tool wraps, unwraps, and strips the wires. To strip insulation from the end of a wire-wrap wire, insert about 1 inch of the wire into the slot on the tool and pull the wire through the slot, as shown in Fig. 1-11. You can also buy convenient prestripped wires in different colors and lengths.

To wrap a connection, insert the stripped end of the wire into the smaller of the two holes in the wrapping end of the tool, slide the larger hole onto the post to be wrapped, and twist the tool several times to wrap the wire onto the post, as shown in Fig. 1-12. To unwrap, slide the unwrapping end of the tool onto the post and twist in the opposite direction.

For components other than ICs, special wire-wrap component sockets can be used or you can wrap connections directly

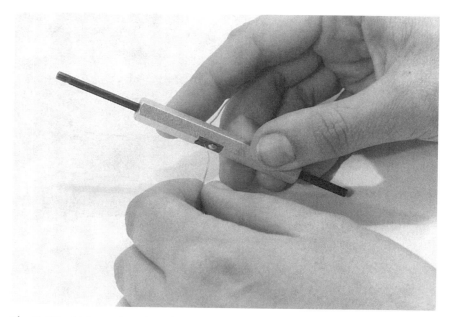

Fig. 1-11 *Using a hand tool to strip a wire-wrap wire.*

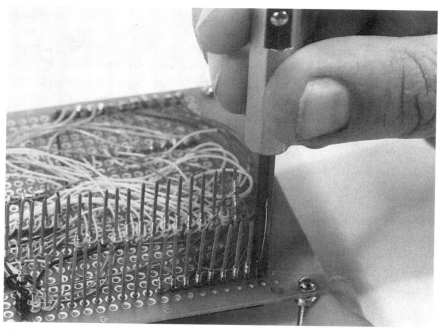

Fig. 1-12 *Wrapping a connection onto a wire-wrap socket.*

onto the component leads. If you wrap onto ordinary, round component leads, it's best to solder the wraps to ensure a good connection.

Like breadboarded circuits, wire-wrapped circuits are easily modified because little or no soldering is involved. The special wire-wrap sockets cost more than ordinary solder-tail sockets, but they are reusable. Wire wrapping is an excellent method for prototyping, especially for circuits that are too large or complex for a breadboard. You can also wire wrap a final version of a circuit, especially if you need just one copy or if you'll want to modify the circuit later.

Wire wrapping can't be used for all circuits, however. It uses thin AWG #30 wire that can safely carry about 100 milliamperes, so high-current circuits are not feasible for wire wrapping. And as with breadboarding, high-frequency circuits aren't practical, and surface-mount components require special sockets. Also, the long socket posts result in bulkier circuits than those built with other methods.

Materials and tools for wire wrapping are widely available from Radio Shack and many mail-order sources. 3M Corporation offers an alternate, low-profile wire-wrapping method called Scotchflex, which requires its own set of tools and sockets.

Point-to-point soldering

A final alternative to pc board construction is point-to-point soldering, where components are inserted into perforated board and connections are made by soldering wires to the component leads as shown in Fig. 1-13. For circuits that are hand-wired, the choice between point-to-point soldering and wire wrapping is largely a matter of personal preference. You can also combine methods; for example, using point-to-point soldering for power supply connections and wire wrapping the others.

Advantages of pc boards

The above methods all have their uses, but compared to other methods of circuit construction, pc boards have several advantages.

Printed circuit boards are easy to duplicate. Once the artwork is created, you can easily make multiple copies of a board. With most fabrication methods, the artwork is transferred onto a board in one operation, instead of having to wire each connection individually. You can even transfer and etch several boards at once

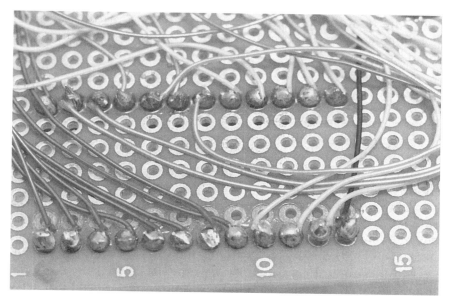

Fig. 1-13 *Point-to-point soldered connections.*

onto a single sheet of base material and cut them apart after etching.

Miswiring is less of a problem with pc boards because you don't have to wire each connection individually. Problems still can occur of course, if the artwork has mistakes or if the artwork transfers or etches imperfectly. You're also not immune to soldering the wrong components on the board or to inserting components backwards, but these types of problems can occur with any construction method. On the whole, mistakes in pc board construction are easier to guard against than having to worry about hand-wiring each and every connection correctly.

Components are easy to locate on the board. On a pc board, each component is installed in its assigned spot on the board. On every identical board, you'll find the same components in precisely the same positions. This makes it easy to spot an incorrect or wrongly oriented component. In addition, pc boards might include component labels and other information such as "pin-1" identifiers for ICs and polarity indicators for diodes and capacitors. These make it easier to insert components correctly and to troubleshoot problems.

Printed circuit boards are neat and durable. You won't find a rat's nest of wiring underneath a pc board as you will with wire-wrapped or point-to-point soldered boards as shown in Fig. 1-14. On a pc board, the connections are part of the board itself (with

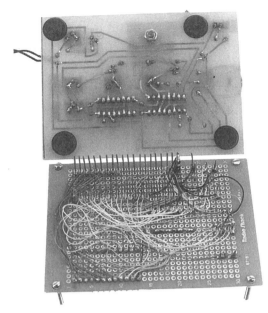

Fig. 1-14   *The solder side of a pc board (top) is much neater and more compact than the underside of a wire-wrapped circuit (bottom).*

the exception of occasional jumper wires). The soldered connections are durable, and there is less danger that jarring or dropping will break a connection.

Printed circuit boards are compact. Compared to wire-wrapped or point-to-point soldered boards, pc boards require only a small amount of clearance, so their enclosures can be smaller.

High-frequency circuits must use pc boards. In circuits that use high-speed logic devices, like those in the emitter-coupled logic (ECL) or advanced Schottky TTL (ASTTL) families, or in radio frequency or microwave frequency analog circuits, the lengths and layout of the interconnections are critical. Printed circuit boards provide the short, fixed-length, fixed-location interconnections these circuits require.

Disadvantages of pc boards

Printed circuit boards do have some disadvantages compared to other methods of circuit construction.

Designing the pc board artwork requires time and skill. With breadboarding, wire wrapping, or point-to-point soldering, you can gather up the components and start wiring right away. A pc board requires the extra step of designing the artwork and transferring it to a pc board before you can construct the circuit. Depending on how complex the circuit is and the kinds of tools you have to work with, this step can take a little time or a lot. If you're in a hurry to get a circuit up and running, a different construction method might be quicker.

Major changes are difficult to make to a pc board without designing a new board. Simple repairs and alterations are easily made, but if you decide to eliminate three ICs and add six others, you're probably going to have to make a new board—although you can reuse the parts of your artwork that are still valid.

If you're using a computer to draw your artwork, making changes is easier because you don't have to redraw from scratch. The software allows you to delete unwanted elements and add others to your original artwork. But you still have to fabricate a new board with the revised artwork.

Artwork mistakes are reproduced on all boards. If you're going to make many copies of a board, it pays to make and test a trial board before you commit yourself to mass production. Any mistakes you make will show up on all of the boards.

Time versus cost

Which method of circuit construction is cheapest or quickest? It depends on the circuit, the available equipment and materials, the number of boards needed, your own skills and experience, and other variables.

Solderless breadboarding is low cost—the only expense is for the breadboard and jumper wires, and these are reusable. Point-to-point soldering and wire wrapping both use inexpensive perforated board as a base material. With wire wrapping, special sockets and other materials add to the expense.

All of the above methods require hand-wiring each connection. This is acceptable for small, one-of-a-kind circuits, but is too time-consuming for larger projects or mass production.

With pc boards, cost depends on the methods used. Applying transfer patterns or using iron-on transfers on a bare copper-coated board are two low-cost methods. Other methods cost more in both materials and equipment, but might be more convenient.

Time is involved in designing the artwork, transferring the image, drilling, and soldering. But the artwork needs to be designed only once, and you can do some of the other steps using batch or assembly-line methods so that the more boards you make, the less time you spend per board.

New developments and challenges

In the world of electronics, nothing stays the same for long. Circuit technologies are advancing rapidly, and pc board technologies are evolving and improving. The following paragraphs describe some of the major developments in these areas.

Surface-mount technology

In the 1970s and 1980s, through-hole components like DIP (dual in-line package) ICs and leaded resistors were popular. Through-hole components are still used, of course, but many designs are switching to new, smaller surface-mount packages. Figure 1-15 shows a comparison of the two packaging technologies.

With through-hole components, the component leads are inserted into drilled holes and soldered to pads on the opposite side of the board. In contrast, surface-mount components rest on one surface of the board.

Generally surface-mount packages are much smaller than equivalent through-hole packages, so they require smaller, more closely spaced pads, thinner traces, and more precise soldering. On the other hand, surface-mount components result in smaller, more compact circuit boards, and eliminate the need to drill lead holes.

Fig. 1-15   *A comparison of through-hole and surface-mount components. Top row: 14-pin ICs, 8-pin ICs, transistors, LEDs. Bottom row: electrolytic capacitors, tantalum capacitors, resistors.*

Large-scale integration

Another advance in semiconductor technology is large-scale integration, where more and more transistors and other circuit elements are packed onto a single chip. This means that modern circuits require fewer components than equivalent circuits using older-technology components. For example, instead of designing a computer circuit with a microprocessor and dozens of memory and peripheral chips, you can now buy a single microcomputer chip with memory and peripheral functions built in. The fewer components there are, the easier it is to design the artwork.

Low power

A circuit built with modern technology might also consume less power than its equivalent built with older technology. For example, in the 1970s, logic circuits were built with transistor-transistor logic (TTL) components, while newer circuits are more likely to use components in the high-speed CMOS (HCMOS) family. Power consumption for an HCMOS chip can be 100 times less than that of an equivalent TTL chip.

Using low-power components means you can use smaller-capacity power supplies with smaller transformers, regulators, and heat sinks. Heat dissipation is less, and the pc boards and components are smaller and less expensive.

High speed

One challenge prompted by new semiconductor technologies is that the new high-speed logic chips require proper board layouts to prevent problems such as ringing, reflections, cross talk, and signal attenuation. Proper component layout and routing of traces can be critical for these circuits.

Computer-aided design

Along with changes in circuit components have come developments in board fabrication. Perhaps most important, personal computers have greatly eased the task of designing pc board artwork. Deciding on component locations and drawing the interconnections is a craft that requires knowledge, patience, and attention to detail. But computers can take much of the drudgery out of the design work.

The advantages begin when you draw the schematic diagram. With special schematic capture software, you can draw a schematic diagram on a computer, store information about the drawing in a special file called a *netlist,* and then import the netlist into a pc board design program. Using the information in the netlist, many pc board design programs can help you place components and route traces in the artwork, as well as check your design when it's finished to find missing connections, spacing violations, or other problems.

When you design artwork with pc board design software, you can draw pads and traces of the exact dimensions you require and place them precisely where you want them in the artwork. You can zoom in to examine your artwork in detail, and zoom out to view the entire board at once. You can draw multiple layers in different colors and view them together or separately. Modifying a design on computer is easy because it's all done from a keyboard, mouse, or other input device, with no tape to lift or add or pencil lines to erase or redraw.

In addition, artwork stored in a computer is easy to print or plot at actual or enlarged scale using a dot matrix, laser, or other printer or pen plotter. Special file formats can be read directly by photoplotters, which plot the artwork onto photographic film, and by automated drilling systems, which drill precise, accurate lead holes in the board.

A design stored on disk is also durable and easily reproduced. Even the most carefully stored artwork can be damaged by dirt, scratches, or tape that loosens and falls off. A file on disk, along with a backup copy, has none of these problems.

New fabrication methods

Another major advance is the development of new and improved fabrication methods for pc boards. New iron-on and direct-plot methods of image transfer make it easier than ever to quickly produce a board. For those who use traditional photographic methods, improved photosensitive board coatings allow for finer traces and use less-toxic developers and strippers.

The future

Continue to watch for new advances in products and methods for pc board design and fabrication. Software for pc board design will continue to become easier to use, with more and better capabilities. New, improved board fabrication methods will continue to be developed as researchers and experimenters search for simpler, faster, safer, and more capable processes.

Working safely

Before getting into the details of pc board design and fabrication techniques, please read and heed the following important information about health, safety, and environmental concerns as they relate to pc boards.

Many of the processes described in this book involve materials or equipment that must be used properly to avoid damaging the health, safety, and environment of you and those around you. Ignoring health and safety precautions can result in injury, death, or pollution to the air you breath or the water you drink.

This book includes cautions and warnings for many of the hazards you might encounter when you carry out the processes described. However, the ultimate responsibility for health and safety is yours. Your projects will proceed more smoothly and safely if you follow these guidelines for health, safety, and the environment.

Read and follow the instructions, warnings, safety precautions, and advisories that are included with the materials and equipment you use. Hazardous chemicals should come with a material safety data sheet (MSDS) that describes the contents and proper handling and disposal of the chemical. If you don't receive one, ask for it. The MSDS includes the chemical name and the manufacturer's name and address, as well as data on hazardous ingredients, physical characteristics, explosion and fire hazards, health hazards, reactivity, spill or leak procedures, personal protection, special precautions, and transportation precautions. Read these forms and keep them handy in case of an accident.

When you have a choice, choose the less-toxic alternative. For example, if you use photographic methods to expose your pc boards, newer aqueous processes use much milder developing and stripping solutions than older solvent-based methods. If your solder contains a flux whose residues are noncorrosive and nonconductive, you're spared the trouble of having to clean your boards with harsh solvents after soldering. Don't cause yourself the hassle of storing, using, and disposing of toxic chemicals if you don't have to.

Store chemicals and tools in a safe, childproof place. Do not store chemicals in the kitchen or anywhere food is stored. Avoid locations that are subject to extreme hot or cold temperatures. Be especially sure that dangerous items are kept away from curious kids.

Keep a separate set of bottles, jars, buckets, and other containers and utensils for use in your pc board projects. Do not store chemicals in food containers or wash and reuse workshop containers or utensils in the kitchen. There's too much potential for mix-ups or contamination.

Clearly label all containers of chemicals. Don't assume that six months or even six days later you will remember what you've stored in which bottle.

Keep your work area clean, uncluttered, and well organized. Avoid accidents caused by tripping over carelessly routed electrical cords or other debris, or knocking over items as you reach through a maze of clutter for the item you want.

Find out and follow local recommendations for disposing of chemicals. Etchants, photographic chemicals, and organic stripping solutions are some of the chemicals that might require special disposal. Many communities sponsor periodic "clean sweep" days where small quantities of toxic chemicals are accepted and safely disposed of. Do not assume that the small quantities you flush down the drain or place in the trash are insignificant.

Protect your eyes. Wear safety goggles when drilling, cutting pc boards or wires, soldering, or performing any activity that could result in particle fragments or chemicals getting into your eyes. Accidents happen quickly and unpredictably. If you use an ultraviolet lamp for exposing artwork, do not look at the light source. Ultraviolet frequencies can cause permanent eye damage. A lamp with a remotely operated timer is the safest option.

Do not wear loose clothing or jewelry that might become tangled while drilling or cutting.

Don't allow fumes to concentrate when developing, etching, soldering, or performing any process that generates fumes. Use an exhaust fan or vent hood to blow fumes away from your work area, preferably to the outdoors.

Wear an appropriate respirator or dust mask to prevent inhaling harmful fumes or fine particles.

Keep a first-aid kit handy in case of accidents.

Always take the time to work carefully and safely. Do not take shortcuts that compromise health or safety.

2
Preparing a useful schematic diagram

Before you can begin to draw the artwork for a pc board, you need a schematic diagram of your circuit. The schematic diagram, or schematic, shows all of the components in the circuit and how they interconnect. Resistors, capacitors, op amps, logic gates, switches, and other components are represented by symbols in the schematic.

A circuit's schematic is distinct from, but closely related to, its pc board artwork. A complete, easy-to-read, and accurate schematic makes the artwork easier to prepare. You can draw a schematic by hand, or with the aid of a computer, or you can use an existing schematic from a book or other source. Two companions to the schematic are the parts list or bill of materials, which lists all of the components in the circuit, and the netlist, which assists pc board design software in drawing the artwork.

This chapter includes a review of schematic diagram basics, tips for drawing a complete, accurate, and easy-to-use schematic and parts list, descriptions of manual and computer-aided schematic drawing techniques, and a buyer's guide to finding schematic drawing software to suit your needs and budget.

Schematic diagram basics

Except for the very simplest designs, you will need a schematic diagram of your circuit before you can begin to design the pc board artwork. The schematic and parts list should contain enough information so that someone familiar with similar circuits could construct or troubleshoot the circuit by referring to the diagram.

In the schematic diagram, symbols represent circuit elements such as resistors, capacitors, logic gates, op amps, and so on. Each symbol has one or more inputs, outputs, or other connecting points. Lines drawn between the connecting points represent the electrical connections between them. A dot at an intersection indicates an electrical connection between two or more lines. Figure 2-1 shows an example schematic.

Most circuits have many connections to power supplies and ground. To keep the schematic cleaner and neater looking, connections to these points are usually indicated by special symbols instead of a maze of interconnecting lines. Signals that connect to many locations, or that are routed off-board, might be labeled by name instead of drawing connecting lines.

Each component in a schematic has a reference designation consisting of a letter or letters and a number. The letter identifies the class of component, and often provides a clue as to the component type. For example, R indicates a resistor, C indicates a capacitor, and so on. The American National Standards Institute (ANSI) publishes a list of recommended reference designations in its standard #Y32.16-1975. The number in the reference designation identifies the particular component within the class. For example, resistors in a schematic are numbered R1, R2, R3, and so on. In addition to the reference designation, where appropriate, components are labeled with a value (10K, 500pF, 300mh), an IC number (74HC00, LM317, 80C51), or other descriptive information.

There are many good books that offer more information on how to draw schematics and design circuits (see appendix C). Manufacturers also publish data books and application notes for their components, and these can be good sources for circuit designs as well. Electronics magazines are another source that contain circuit ideas you can use or adapt to your own needs.

Guidelines for an easy-to-use schematic

Following some basic guidelines when drawing your schematic will make the schematic easier to understand and work with as you create your pc board artwork and construct and test your final circuit. Here are some guidelines to consider as you draw your schematics.

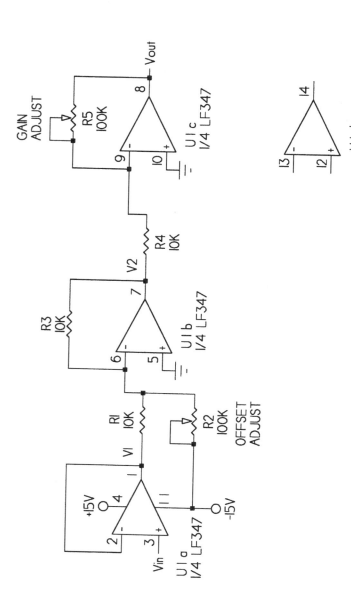

Fig. 2-1 *A simple schematic diagram.*

Group related components together. All but the simplest circuits can be divided into functional blocks, where groups of components work together to accomplish particular functions or tasks. Grouping related components together on the schematic makes the circuit easier to understand, lay out on a pc board, and troubleshoot.

In general, maintain signal flow from left to right on the schematic. Place inputs on the left and outputs on the right. Following a standard, predictable signal flow makes it easier to trace a signal through the circuit.

When possible, group together all connections that are routed off the pc board. Many pc boards connect to wires or connectors that are routed off-board to cable connectors, a backplane or motherboard, or other connections. Grouping the off-board connections together on the schematic, usually along one side of the sheet, makes them easy to find.

When a schematic covers multiple pages, indicate the origins or destinations of off-page signals by page number and location on the page. This is a convenience to anyone who needs to trace a signal through the schematic.

As much as possible, avoid intersecting lines that do not connect electrically. A connection is easiest to trace visually if it crosses only those lines to which it connects electrically. When you have a maze of intersecting lines, some connecting and others not, you have to work harder to trace the connections. Figure 2-2 shows ways to redraw a schematic to eliminate intersections of nonconnecting lines. Some crossed lines might be unavoidable. Be sure to use connect dots to indicate those lines that connect electrically.

For clarity in computer circuits, use a single thick line to indicate a bus. Computer circuits typically contain buses, which are signals that are routed as a group to computer chips, memory ICs, or other components. For example, a data bus might consist of eight signal lines, each of which carries one bit of data. Other common buses hold address or control signals. In the schematic, you can draw a separate line for each of the signals, but this often results in a cluttered and confusing maze of lines on the page. An alternative is to use a single, thick line to indicate the entire bus, with individual signals branching from it. Each line that branches from the bus must be labeled. Figures 2-3 and 2-4 show circuits drawn without and with a data bus.

Number all ICs, resistors, capacitors, switches, and other components. For example, resistors might be numbered from R1

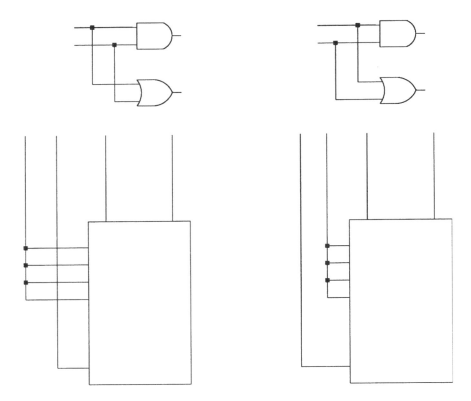

Fig. 2-2 *Two examples of how to redraw a schematic to eliminate unnecessary crossed lines.*

through R22, ICs from U1 through U4, and so on. When a single IC contains multiple gates, op amps, or other units, give each a unique label (U1a, U1b, for example). Numbering should follow a general pattern of left to right and top to bottom on the page, following the signal flow of the circuit, so that components are easy to locate on the page by number.

Indicate polarity for electrolytic capacitors and other polarized components. This is a convenience for those building or testing the circuit.

If an IC's power supply and ground pins have unusual locations include them in the schematic. On many ICs, the power supply and ground pins are in the traditional, expected locations; for example, on most 14-pin logic ICs, pin 7 is ground and pin 14 is V+. For these ICs, showing the power supply and ground pins on the schematic is optional. But other ICs don't follow the standard locations. For these components, include the power supply and ground pin connections somewhere on the schematic. You can draw them directly on the component or list them separately, as shown in Fig. 2-5.

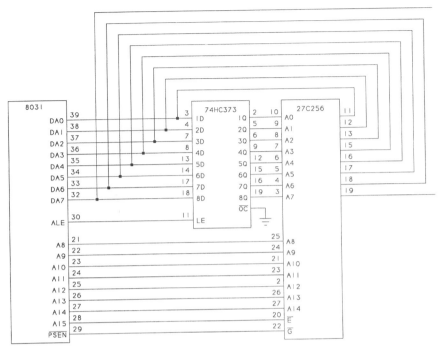

Fig. 2-3   *This schematic does not use buses, instead, each connection is drawn separately.*

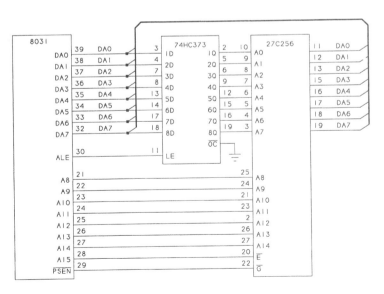

Fig. 2-4   *Fig. 2-3 is redrawn here using a bus, replacing eight signal lines with one thick line.*

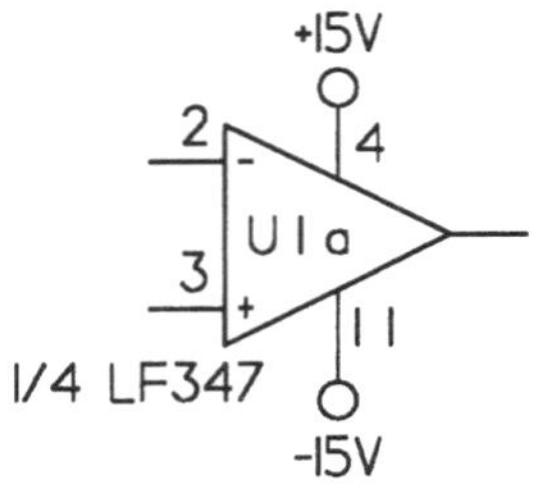

Power supply and ground pins

Chip	+5V	GND
U3	40	20
U4	28	14
U8	16	15

Fig. 2-5 *Two ways to indicate power and ground pins on a schematic.*

Include a labeled sketch for unfamiliar component packages. This is another aid to the board designer or troubleshooter. For many component packages, the leads are numbered in a standard way, beginning at a "pin-1" dot or notch on the component. If the pin numbers are included in the schematic, users can easily match the component pins to the pin numbers on the schematic.

Pin designations for other components are less obvious. For example, all bipolar transistors have a base, emitter, and collector, but not all have pins in the same order on their packages: one might be B-C-E while another might be E-B-C. By including a sketch of unfamiliar component packages, you'll save time and trouble should you ever need to test or troubleshoot the component on the board. Figure 2-6 shows an example.

Label any special fabrication requirements. This includes components or leads that have special mounting or layout requirements, such as heat sinks, heat-sensitive components that should be kept away from heat-generating components, shielding, special spacing requirements, or mounting hardware for heavy components.

Indicate decoupling capacitors. Large circuits might have a dozen or more power supply decoupling capacitors spaced evenly across the pc board. Don't forget to include these on the schematic. You can draw each individually, or draw just one and give it a multiple designation (C12–C16, for example), as shown in Fig. 2-7.

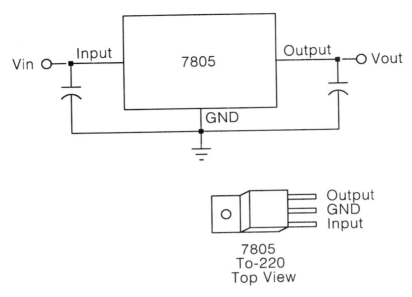

Fig. 2-6   *Draw a labeled sketch of component packages that might be unfamiliar to users of the schematic.*

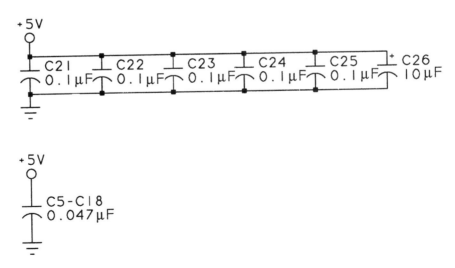

Fig. 2-7   *Two ways to indicate decoupling capacitors in a schematic.*

If a circuit contains unused logic gates, op amps, or other circuit elements, draw these in the schematic as well. For example, if a circuit uses three out of four gates in a quad logic IC, draw the unused gate and label its pins on the schematic. This way, users will know the gate is there if it's needed later.

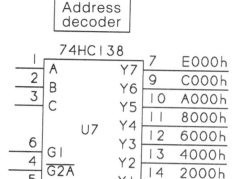

Fig. 2-8 *Label the functions of components and signals in your schematic to help users understand how the circuit works.*

If a circuit is complex, and especially if it covers multiple pages, consider drawing a block diagram that shows an overview of the basic circuit elements including a list of pages where the schematic for each can be found, if appropriate.

Label unfamiliar ICs by name as well as by number, and label functional blocks in the circuit. For some ICs, such as logic gates, the basic function is obvious by the shape of the symbol. The functions of other ICs are less clear, and it can be very helpful to have component labels to clarify functions. When a group of components performs a single function in a circuit, you can label this as well, as Fig. 2-8 illustrates. Even if you will be the only one to use the circuit, you might find that a month after designing the circuit, the operating details are no longer fresh in your mind, and reminders like these are helpful.

Indicate test points. A test point can be any point in the circuit where you might monitor a signal with an oscilloscope, voltmeter, or other test instrument. Test points can be very helpful when you examine or troubleshoot the circuit. Common test points include amplifier and digital logic block outputs. You can add buffered or unbuffered terminal pins to the pc board to act as test points, or you can simply monitor component pins where these are accessible. In addition to labeling the test points on the schematic, you might describe or sketch the expected signal (Fig. 2-9).

Check your work. After your schematic is drawn, but before you begin to design your pc board artwork, it pays to double-check your work to ensure that your schematic is complete and accurate.

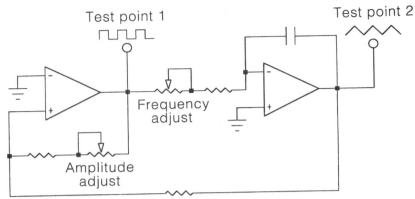

Fig. 2-9   *Help users of your schematics by indicating test points and expected signals and signal levels, where appropriate.*

Is it complete? Double-check that no components or connections have been forgotten. Schematic drawing software that includes design checking can help you find errors and omissions.

Is it accurate? Double-check all IC pin assignments for accuracy. Verify that component values (resistors, capacitors, etc.) and part numbers are correct. It's easier to correct mistakes like these now, rather than after the artwork is designed or the circuit is built.

The parts list

The parts list describes the components used by a circuit. A bill of materials is a complete parts list that might include vendors and prices as well as part descriptions. The parts descriptions should include full part numbers or other descriptive information. For example, instead of 390-ohm resistor, say 390-ohm resistor, through-hole, ¼-watt, 5% tolerance. To simplify things, you can specify: All resistors are through-hole, ¼-watt, 5% tolerance unless stated otherwise. Figure 2-10 is an example parts list.

Schematic drawing software can help you prepare the parts list by automatically generating a list of components used in the drawing. You can also prepare your list manually, on paper or with a text editor or word processor. Either way, double-check to be sure the list is complete and accurate.

```
|--------------------------------------------------------------------------|
|Bill Of Materials for LOGICPRX.SCH on Sat Jul 25 09:18:27 1992            |
|--------------------------------------------------------------------------|
|Item|Qty|Reference|Part Name    |Package   |Description                  |
|----+---+---------+-------------+----------+-----------------------------|
|1   |1  |C1       |CAP,0.1uf    |DCAP\SR21 |DECOUP CAP RADIAL BODY:.270  |
|    |   |         |             |          |X.130 CENTERS:.200           |
|2   |4  |CHA CHB  |CON\SIP\1P   |SIP\1P    | GENERIC 1 PIN CONNECTOR     |
|    |   |GND V+   |             |          |(PROBE)                      |
|3   |1  |D1       |DIODE        |DO7       |Axial ss Diode w Alternate.  |
|4   |4  |LED1-4   |LED          |LED       | LIGHT EMITTING DIODE        |
|5   |1  |U1       |LM339        |DIP14     |Quad Comparator Gate         |
|    |   |         |             |          |Swappable                    |
|6   |2  |R4 R12   |RES,10k      |R1/4W     |RES BODY:60   CENTERS:400    |
|7   |2  |R1 R9    |RES,120k     |R1/4w     |RES BODY:60   CENTERS:400    |
|8   |2  |R8 R16   |RES,150k     |R1/4w     |RES BODY:60   CENTERS:400    |
|9   |2  |R6 R14   |RES,200K     |R1/4w     |RES BODY:60   CENTERS:400    |
|10  |4  |R2 R5 R10|RES,200k     |R1/4w     |RES BODY:60   CENTERS:400    |
|    |   |R13      |             |          |                             |
|11  |4  |R3 R7 R11|RES,470      |R1/4w     |RES BODY:60   CENTERS:400    |
|    |   |R15      |             |          |                             |
|--------------------------------------------------------------------------|
```

Fig. 2-10   *An example parts list, or bill of materials.*

Drawing technologies

When you are ready to draw your schematic, you have a choice between drawing by hand or using computer-aided design (CAD, or CADD, for computer-aided drafting and design). Computer-aided schematic drawing systems range from basic systems that do little more than help you draw lines and text and print and store your drawings, to full-featured products with extensive component libraries, design checking, automatic generation of a bill of materials and other specialized files, and many other features.

You can also use a computer to draw the pc board artwork. Sometimes a single software product can handle both schematics and pc board artwork, while other products are limited to one of these functions. Some companies offer separate but compatible products for schematic drawing and pc board design.

Computer technology continues to advance rapidly. Every year you can do more for less money. The capabilities of CAD systems that once cost tens of thousands of dollars are now available in personal computers for a fraction of the price. Products are available to fit any budget.

Schematic drawing by hand

The development of computer-aided drafting doesn't mean that hand-drawn schematics are obsolete. For years, schematics were

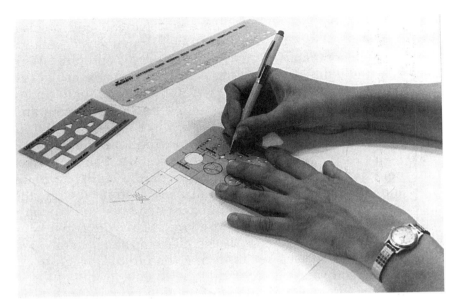

Fig. 2-11 *Use templates for neat hand-drawn schematics.*

drawn without computers, and you can still draw a schematic diagram this way. Figure 2-11 shows a schematic being drawn with the aid of a *template,* a sheet of tinted plastic with cutouts of commonly used symbols for easy tracing onto paper. Templates are available with schematic symbols, basic geometric shapes, and numbers and letters for labeling.

For a durable, professional-looking, hand-drawn schematic, use a mechanical drafting pencil with hard or medium lead and good-quality drafting paper. These are available at art supply or engineering supply stores.

Computer-aided schematic drawing

With computer-aided schematic drawing, you use a keyboard, mouse, or other input device to draw the schematic on a computer screen. The schematic can then be stored on disk, where you can load it into memory for viewing, editing, or printing. Some schematic drawing software goes beyond schematic drawing and also stores information about the schematic for later use in pc board design. Software with this capability is called *schematic capture* software.

Computer-aided schematic drawing has several advantages over manual methods. With a computer, you can effortlessly draw perfect straight, parallel, perpendicular, or curved lines, as well as consistent, repeatable shapes and symbols and precise, neat text. And like other computer-aided drawings, schematics drawn with a computer are easy to modify, copy, and print. With most schematic drawing programs, you can easily exchange one component for another, move a group of components to a new location in the drawing, or remove and add components and connections.

Software suitable for schematic drawing is available in all price levels and in many levels of sophistication. The only absolute requirement for schematic drawing is the ability to place lines and text in a drawing. Other important features include the ability to store and retrieve component symbols and to copy, move, or delete portions of the drawing. Many programs have additional capabilities such as design checking and generation of a netlist, which is used in importing the circuit design to pc board layout software.

Software for schematic drawing includes products that are intended specifically for schematic drawing, as well as general-purpose CAD programs that you can use for mechanical, architectural, or other drawing.

Painting versus CAD software

Drawing software for computers is available in two basic types: paint programs, which use raster-based, or bitmapped, graphics, and CAD programs, which use vector-based graphics. For schematic drawing and especially for drawing pc board artwork, CAD software is best because it can create the most detailed, precise, and accurate drawings. This is especially important when drawing pc board artwork that contains narrow traces and tiny pads that must be placed precisely and accurately.

In schematic drawing, precision is less critical, but this introduction to computer-aided design is a good place to learn about the differences between painting and CAD software. With either type of software, you create drawings using a keyboard, mouse, or other input device to place and draw lines and objects on screen. But CAD and paint programs store their drawings in different ways. To understand the difference requires knowing something about how computer video displays and printers operate.

A typical video screen contains thousands of *pixels*, or picture elements, arranged in a grid of rows and columns. A pixel that is turned on appears as a bright dot of light, while a pixel that is off is dark. A line drawn on screen is really a series of dots placed so closely together that they appear to be a continuous line. Colors are generated by turning on combinations of colored dots at the same location.

High-resolution displays have more pixels than low-resolution displays, resulting in higher-quality pictures. For example, IBM's low-resolution CGA display has a grid of 640 by 200 pixels, while the higher-resolution VGA display has 640 by 480 pixels. A circle drawn on a high-resolution display is smoother and less jagged than the same circle drawn on a display with lower resolution.

Like video displays, printers also draw with dots and vary in their resolutions. A printer that prints 300 dots per inch can draw smoother curves than one that prints only 100 dots per inch. High-resolution printers are also capable of drawing thinner, more closely spaced lines.

Paint programs store shapes and drawings as a screen image, or as a grid of dots as they appeared on the screen. This means that a drawing created with a paint program is limited to the resolution at which it was created. Even if you print the drawing on a high-resolution printer, it will have no more detail than it did on screen.

In contrast, drawings created with CAD software don't have this limitation. In a CAD drawing, lines and other shapes are stored as objects, or idealized mathematical descriptions of the shapes. For example, a circle is stored as the location of its center in the drawing and the length of its radius. When the drawing is displayed or printed, the video screen or printer displays or prints the circle as best as it can, using the stored information to determine its location and size. This means that even if you create a drawing on a low-resolution screen, you can generate a high-quality printout on a high-resolution printer. Figure 2-12 illustrates the difference between the two types of software.

For the most precise, detailed, highest-quality drawings, choose CAD software over a paint program. Most software designed for pc board work is CAD software.

Step-by-step: Drawing a schematic with PADS-Logic

PADS-Logic from PADS Software is an example of dedicated schematic drawing software. In addition to a professional version of the software, PADS Software offers a free evaluation copy that includes the same features as the professional version, with

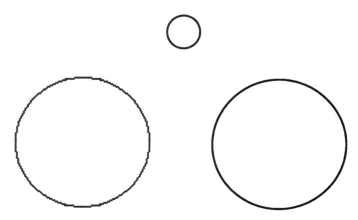

Fig. 2-12 *A small circle drawn on-screen using raster-based software will look "jagged" when printed at a larger size, while the same circle drawn with vector-based software retains its smooth curve when enlarged.*

the limitation that drawings can contain no more than about 30 ICs. This makes the software fine for use on many simpler circuits or as evaluation software if you are considering buying the professional version.

The following sections describe PADS-Logic in greater detail. PADS-Logic is just one example of schematic capture software. There are many other products that perform the same basic functions. They vary in how they operate and in the additional features they offer.

Figure 2-13 shows an example PADS-Logic screen. The screen is divided into four parts.

The *working area* takes up most of the screen. The drawing grid is defined by the evenly spaced dots across the working area.

The *system information window* is in the upper left corner of the screen. It contains the current cursor position in relation to the drawing origin, the drawing name, the current sheet number and the number of sheets the drawing uses, the grid size, a "postage stamp" window that shows how much of the current sheet is displayed, and the current menu path. Often the screen shows only a small portion of a drawing or sheet, and the postage-stamp window shows you which part of the sheet you are viewing.

The *command menu* is below the system information window. The main command menu consists of several items. Some of the items call up submenus with additional functions.

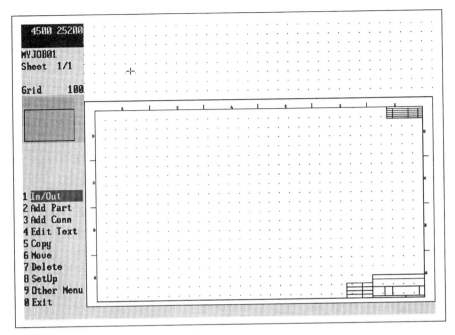

Fig. 2-13 *A PADS-Logic drawing screen.*

The *prompt line* is along the bottom of the screen. It displays messages and keyboard input as appropriate.

The functions of the items in the main command menu include

- *Job In/Out.* For loading a drawing, printing and plotting, and generating netlists, bills of materials, and other reports.
- *Add Part.* For placing components in the drawing.
- *Add Connection.* For drawing lines to connect components.
- *Edit Text.* For changing part designations.
- *Copy, Move,* and *Delete.* For copying, moving, and deleting items or groups of items.
- *SetUp.* For selecting display colors and preferences.
- *Other.* Miscellaneous functions, including adding text, adding a bus, and defining a group of items.

You can select items from the command menu with their designated function keys or by moving the cursor into the command menu area and clicking the mouse.

From the numeric keypad, you can zoom in or out to view any area, from the entire drawing to any section you select with the cursor.

The following steps demonstrate how to draw a schematic using PADS-Logic. The procedure described below is one that has worked for me. You might have different preferences about grid size, order of parts placement, and other details. The procedure assumes that your software has been installed and configured to match your hardware and preferences. It also assumes you are familiar with basic computer use.

1. Have a sketch of your schematic handy, or if you prefer to draw directly on screen, have available the design requirements, data sheets, and other information you'll need to design the schematic.
2. Run PADS-Logic. This calls up the main screen and displays the working area and main menu.
3. Zoom in to view an area of about 5 by 6 inches. This will show enough detail so that you can place components precisely in the drawing. Zoom by pressing *Del* on the numeric keypad, then draw a box with the cursor to define the area you want to view. Pressing *Home* returns you to the full drawing. You can change the size or location of the viewing area whenever you want.
4. Begin placing major components in the drawing (Fig. 2-14). To place a component, select *Add Part* from the main menu, and type a component name on the command line. The screen will show an outline of the symbol selected (Fig. 2-15). Select *Accept* in the *Add Part* menu, and place the component in the drawing with the cursor. Using your sketch as a guide, try to place the components more or less where they will be in the final drawing. Don't worry too much about precise placement because you can move components later if you wish.
5. Add connections as required (Fig. 2-16). Select *Add Conn* from the main menu, place the cursor on the first node to connect, and click the mouse. As you move the cursor, a line will follow, stretching from the previous cursor position to its current position. To make a corner, click the mouse. When the line stretches to another node, the connection is complete. Selecting *Complete* from the menu completes the connection on screen.
6. Browse for components as necessary. If you're not sure of a component's name, you can browse for it in the compo-

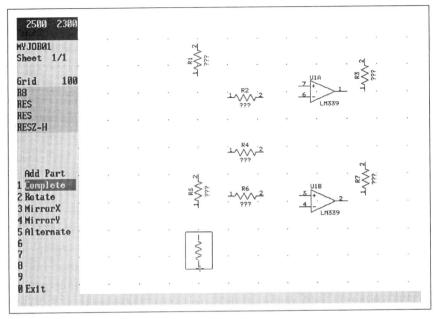

Fig. 2-14 *Adding components to a drawing.*

nent library. For example, to select a capacitor symbol, select *Add Part* from the main menu and type *C** to see a list of part names that begin with C. Use the arrow keys to see an outline of each symbol in turn. When the symbol you want is highlighted, select *Accept* and place it in the drawing. Some components have alternate symbols, which you can select with *Alternate* in the *Add Comp* menu (Fig. 2-17).

7. Add power and ground connections. Select *Add Conn* from the main menu, select *Ground* or *Power* as appropriate, and place the symbols in the drawing. Alternate symbols are available for these as well.

8. Add component designations. Select *Edit Text* from the main menu and use the cursor to select the text to edit (Fig. 2-18). If you prefer, the *SetUp* menu allows you to configure the software so that it prompts you for component designations as you place each component.

9. Move, copy, delete, and change components and connections as necessary. As your drawing evolves, you'll probably want to make some changes, rearranging components and connecting lines for a neater, clearer drawing. The main menu's *Move, Copy,* and *Delete* items perform these

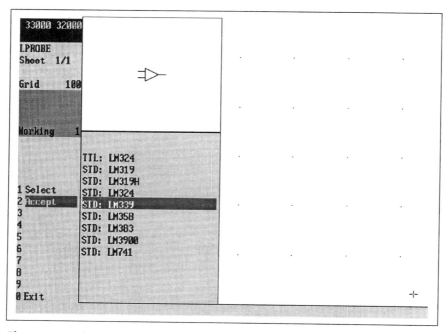

Fig. 2-15 *Selecting a component from the component library.*

functions. As Fig. 2-19 shows, when you move a symbol, its connecting lines follow along with it. You can swap pin locations in a symbol by selecting *Edit Text* and the two locations to be swapped. The swap occurs automatically. The *Add Part* menu allows you to rotate or mirror a symbol. Fig. 2-20 illustrates the group copy function.

10. Save the drawing. Select *In/Out* from the main menu, then *Job Out,* and type a file name to store your drawing.

11. Print the drawing. You can print the entire sheet or only a portion of it. To print a portion, zoom in to display the area you want to print. Select *Job In/Out,* then *Plot.* A list of printing options is displayed. To position the printout on the sheet, select *Scale to Fit,* which causes the drawing to fill the sheet. When the options are correct, select *Complete,* and the printer will print the schematic.

Creating a netlist

A netlist provides the key to translating a schematic into pc board artwork. A *net* is a connection between two or more components, or between a component and a cable connector or

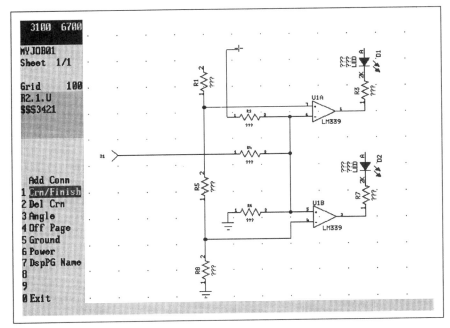

Fig. 2-16 *Adding connections in a drawing.*

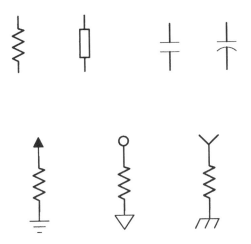

Fig. 2-17 *Alternate schematic symbols. Top row: resistors, capacitors. Bottom row: Three resistors that connect to alternate power supply (top) and ground (bottom) symbols.*

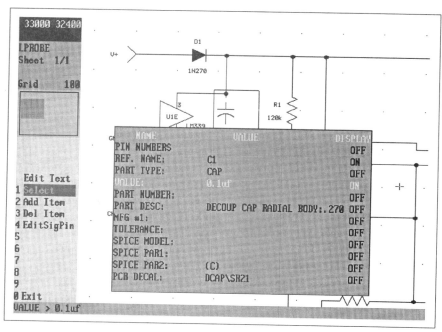

Fig. 2-18 *Adding component designations to a drawing.*

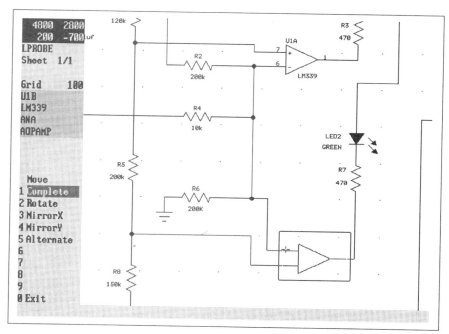

Fig. 2-19 *The op amp surrounded by a box is in the process of being moved. As the component moves, its connecting lines stretch or shrink to fit as needed.*

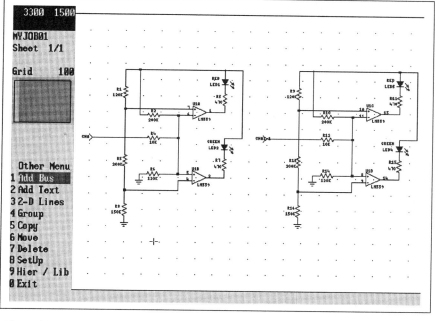

Fig. 2-20 *Group copy reproduces a group of components and their connections.*

another off-board connection. As you might guess, a netlist is a list of the nets in a schematic diagram.

Figure 2-21 shows an example netlist in PADS-PCB format. The netlist begins with a list of the components used. Following this is a list of the nets, or connections, in the drawing. Each net is described by type, net name (assigned by the software), trace width in mils (1 mil = 0.001 inch), and the pins to which it connects. Some nets connect just two pins, while others have multiple pins that interconnect.

Many pc board layout programs can read and interpret netlists. The netlist tells the program what components a circuit contains and how they connect so you don't have to reenter the information manually when you design the artwork. Often the netlist also includes information about component sizes and packages and trace widths for use in creating the pc board artwork.

In PADS-Logic, you create a netlist by selecting *Job In/Out,* then *Reports,* then *Netlist.* You choose a format and file name and the netlist is created for you. The netlist is a text file. You can display it from within PADS-Logic or with any text editor.

```
*PADS-PCB* FILE INFORMATION
*REMARK* C:\PADSDEMO\LOGFIL\LOGICPRX.SCH
--ALL SHEETS --
Sat Jul 25 09:18:01 1992

*PART*          ITEMS
U1              LM339@DIP14
R1              RES,120k@R1/4w
R2              RES,200k@R1/4w
R3              RES,470@R1/4w
R4              RES,10k@R1/8W
R5              RES,200k@R1/4w
R6              RES,200K@R1/4w
R7              RES,470@R1/4w
R8              RES,150k@R1/4w
R9              RES,120k@R1/4w
LED2            LED@LED
CHA             CON\SIP\1P@SIP\1P
LED1            LED@LED
R10             RES,200k@R1/4w
R11             RES,470@R1/4w
R12             RES,10k@R1/8W
R13             RES,200k@R1/4w
R14             RES,200K@R1/4w
R15             RES,470@R1/4w
R16             RES,150k@R1/4w
C1              CAP,0.1uf@DCAP\SR21
LED3            LED@LED
LED4            LED@LED
CHB             CON\SIP\1P@SIP\1P
GND             CON\SIP\1P@SIP\1P
D1              DIODE@DO7
V+              CON\SIP\1P@SIP\1P
*NET*
*SIGNAL* $$$5227 13
U1.13 R15.1
*SIGNAL* $$$5229 13
R14.2 R12.2 U1.11 R10.2 U1.8
*SIGNAL* $$$3393 13
CHA.1 R4.1
*SIGNAL* $$$5231 13
U1.14 R11.1
*SIGNAL* $$$5233 13
R9.1 R13.2 U1.9
*SIGNAL* $$$5235 13
R16.2 R13.1 U1.10
```

Fig. 2-21 *An example netlist.*

```
*SIGNAL* GND 50
R6.1 R8.1 R14.1 R16.1 C1.2
GND.1 U1.12
*SIGNAL* $$$5243 13
R6.2 R4.2 U1.5 R2.2 U1.6
*SIGNAL* $$$5245 13
R5.2 R1.1 U1.7
*SIGNAL* $$$5247 13
R5.1 R8.2 U1.4
*SIGNAL* $$$5249 13
R7.1 U1.2
*SIGNAL* $$$5252 13
R1.2 R2.1 LED1.A LED2.A D1.K
C1.1 U1.3
*SIGNAL* $$$5254 13
U1.1 R3.1
*SIGNAL* $$$3453 13
R12.1 CHB.1
*SIGNAL* $$$3456 13
R9.2 R10.1 LED3.A LED4.A
*SIGNAL* $$$3435 13
V+.1 D1.A
*END*       OF ASCII OUTPUT FILE
```

Fig. 2-21 (cont.) *An example netlist.*

There is no standard netlist format, and many schematic capture programs can create netlists in a variety of formats, and many pc board layout programs can read a variety of netlists. You can usually find a format that is compatible with both programs, even if the software is from different companies.

Buyer's guide to schematic capture software

PADS-Logic is just one example of schematic capture software. Products are available from many companies, many of which are listed in appendix A. Computer and electronics magazines occasionally review schematic capture software and also contain ads for new products, so these are good places to look for information about the latest offerings.

If you are in the market for schematic capture software, the following information will help you choose a product that can not only help you draw your schematics, but can help you translate your schematic into a working pc board. Some of the considerations, such as system requirements, are also valid when choosing software for pc board design, which is covered in chapter 5. Other considerations are unique to schematic capture software.

System requirements

When you are choosing schematic drawing software, you want to be sure that your computer hardware meets the software's minimum system requirements. If you are buying a new computer, you can select the computer and schematic drawing software at the same time. Questions to ask about system requirements include the following:

- What type of computer does the software run on (IBM PC, Macintosh)?
- What operating system (MS-DOS, OS/2) or environment (Windows) does it require?
- How much memory is required?
- How much disk space is required?
- What video modes (CGA, EGA, VGA) does it support? The higher the video resolution, the more detailed your on-screen drawings will be. The screen resolution of CAD programs doesn't affect the resolution of the printed or plotted drawings, but it does affect how easy it is to view and work on your drawing on the screen.
- Is a math coprocessor required, recommended, or not needed? A math coprocessor can speed up the operation of CAD programs dramatically, if the software takes advantage of it.
- Is a mouse or other input device required or recommended? Mice are very handy for pointing, selecting, and placing objects in a drawing.
- What printers, plotters, or other output devices are supported?

Placing symbols and lines

All schematic capture programs allow you to precisely place component symbols and lines in a drawing. Typically, you draw a line by selecting the desired beginning and end points with a cursor. You place a component by selecting it from the component library and positioning it in the drawing with a cursor. But different products vary greatly in their ease of use and the convenience of these functions. Questions to ask include

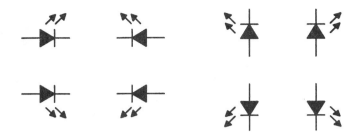

Fig. 2-22   *Different orientations of a component.*

- What grid sizes are available? For precise line placement, with no gaps or misaligned lines, most schematic capture software draws lines and components on a grid, which is a set of evenly spaced horizontal and vertical lines. The grid is usually indicated by small dots at the grid's intersections. Some programs limit you to predefined grid sizes, while others allow you to define your own grid and change or ignore the grid when you wish.
- Can I place components in different orientations (0, 90, 180, and 270 degrees)? Figure 2-22 shows examples.
- Is rubber-banding available? When you draw a line with rubber-banding on, a visible line stretches from the first point selected to the current cursor position. This allows you to see how a line will route before you complete the line by selecting the second point. Without rubber-banding, the line is invisible until you select both end points. When you move a component with rubber-banding on, the connecting lines stretch or shrink as needed to match the new position.

Component libraries

Most programs include, or have available at extra cost, a library or libraries of components that you might use in your schematics. Each component has a unique name. Sometimes one physical component contains multiple logic gates, op amps, flip-flops, or other units. Some programs treat each of these as a separate component (7400a, 7400b, 7400c, 7400d), while others use a single component name (7400) from which you select one when you place it in the drawing. The following questions should be asked about component libraries:

- Does the library contain most of the components you need? No component library can include every available

component, because new ones are constantly being re-
leased. Vendors try to include the most popular devices,
including logic ICs from the major families, and microprocessor components, including memory and peripheral
ICs. The selection of linear devices like op amps, timers,
and voltage regulators is usually more limited, and worth
checking on if these are important to you. In addition,
many devices might be offered in a choice of package types
(DIP, SOIC, PLCC, and so on). The different package types
might have different pinouts, and even if they don't, a
package designation might be needed in the board layout
stage.

- How many components can the library hold? The ideal
  component library has every component you need, but no
  more. The size of the component libraries are limited, by
  disk space if nothing else, but the limit might be in the
  thousands. If necessary, you can create multiple libraries
  for different types of projects. Smaller libraries are easier
  and quicker to search through.

- Can you remove unwanted components? What if the
  library includes a complete set of standard TTL components, and you use only HCMOS parts? Most programs
  allow you to remove unwanted components from a library
  to save disk space.

- Is it easy to design your own components? No matter how
  complete the component libraries are, there will be times
  when you need a component that is not included. For this
  situation, you need a way to create your own components.
  Programs vary widely in how easy it is to do this. Some require you to remember and strictly follow precise design
  rules. Others provide menus, visual cues, and other help
  that makes it easy to draw new symbols, either from
  scratch or by modifying and renaming an existing symbol.
  Figure 2-23 is an example of creating a component with
  PADS-Logic. Some IC vendors even design and release
  component symbols for their new devices for use with
  popular programs, saving you the trouble of designing
  them yourself.

- Is it easy to select a component from a library? Some programs allow you to browse through the component library
  and view each symbol before selecting one. With others,
  you select components by name from a list, and you don't
  see the symbol until you place it in the drawing.

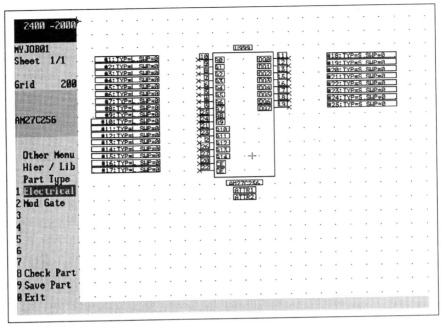

Fig. 2-23 *Designing a new component with PADS-LOGIC.*

Placing text

All schematics must include text for part designations, pin numbers, and other information. The text doesn't have to be typeset quality, but it should be readable and easy to enter and edit. Questions you should ask include

- Is the text size variable? You'll want to use a consistent size of text for parts labeling and other common functions, but being able to change the size of the characters when you want to is convenient.
- Can the text be rotated? A convenient, but not essential, feature is the ability to place text at a 90-degree angle on the page.
- Does the software automatically prompt for component numbers and designations as components are placed? This saves you the trouble of having to add this information in a separate step. Many programs will number components automatically as you place them (R1, R2, and so on).
- Is the text easy to edit? The ability to easily change component values and other text is important.

Editing tools

Putting together a schematic diagram usually involves some trial and error. You might place a group of components and later decide you want to move them to a new location on the page. Or you might want to erase, copy, replace, or otherwise change parts of the diagram. With the right editing tools, functions like these are easy, and certainly much more convenient than with pencil and eraser. Ask the following questions:

- Can I move, rotate, mirror, and copy components and blocks of components within a drawing? Most programs include these basic functions. In some programs, all connections remain, even after moving, rotating, or mirroring a component. You might have to do some editing to clean up the appearance of the drawing, but you don't have to completely erase the old lines and redraw new ones.
- Can I swap components, gates, or pins? Design changes or performance problems might cause you to swap one component for another. For example, you might decide to replace a general-purpose LM741 op amp with an LF351 wide-bandwidth op amp; or in a 74LS00 quad NAND gate, you might want to experiment with different gate or pin assignments to make board routing easier. Some programs make it easy to swap parts, gates, or pins in a single operation, as shown in Fig. 2-24.
- Does the program place borders and title blocks automatically? Figure 2-25 shows an example title block. If the software doesn't create one automatically, you can design your own.
- Will the program locate a component on screen by name? In large, complex schematics, this feature allows you to name a component and ask the program to highlight each of its locations (if any) on the schematic.
- Does the program perform design checks? Some programs will check through your schematic on request, looking for and reporting design flaws like duplicate component designations, open pins that should be connected, and other problems.

Screen functions

Screen functions have to do with how you control a drawing's appearance on screen.

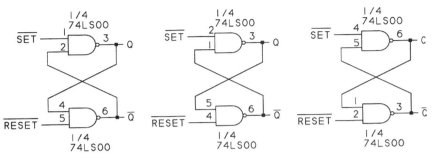

Fig. 2-24 *Sometimes swapping pins (middle) or gates (right) can make pc board layout easier.*

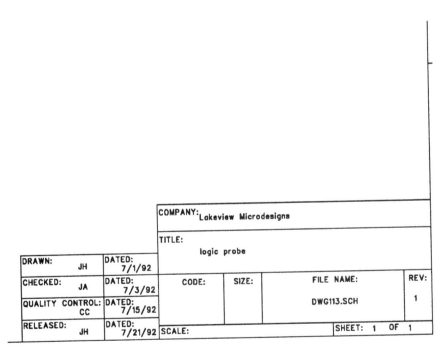

COMPANY: Lakeview Microdesigns						
TITLE: logic probe						
DRAWN: JH	DATED: 7/1/92					
CHECKED: JA	DATED: 7/3/92	CODE:	SIZE:	FILE NAME:		REV:
QUALITY CONTROL: CC	DATED: 7/15/92			DWG113.SCH		1
RELEASED: JH	DATED: 7/21/92	SCALE:			SHEET: 1 OF 1	

Fig. 2-25 *A title block helps to document your schematic.*

- Can I zoom in and out to view a small section or the entire drawing at once? Are zoom scales fixed (×1, ×2, etc.) or user-definable? Figure 2-26 shows a screen that is zoomed in to show one small part of a schematic in detail.
- Can I pan around the drawing to view different sections of the drawing? Some programs have autopanning, which automatically shifts the display left, right, up, or down whenever the cursor reaches the edge of the screen.

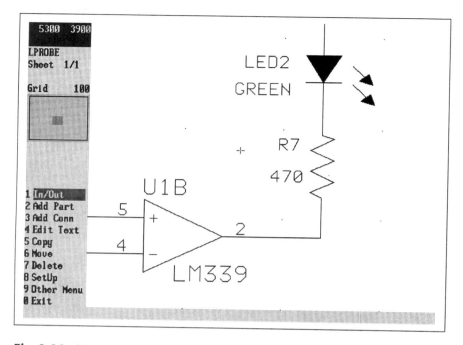

Fig. 2-26 *Most programs allow you to zoom in or out to view a small portion of a drawing or the entire drawing at once.*

- Can I view multiple windows or drawings simultaneously? Sometimes you might want to see two different schematics, or zoom in on two or more areas within a schematic, at the same time. Splitting the screen into multiple windows allows this.

Schematic limits

Be sure that any program you buy can handle the largest, most complex schematic you intend to draw.

- What physical drawing sizes are allowed? Some programs will automatically fit a drawing to a standard-size sheet, corresponding to the sizes of standard drafting paper sizes.
- Can a single schematic be drawn over multiple sheets? Some programs allow a drawing to cover several sheets. Signals with the same name that appear on more than one sheet are automatically connected in the netlist. Other programs limit you to a single sheet per schematic.
- What is the maximum number of ICs, connections, and components per sheet and per schematic? Be sure the software can handle the biggest circuits you expect to draw.

User interface

The user interface refers to how you interact with the software.

- How many screen colors are available? Are screen colors user selectable? For schematic diagrams, colors aren't essential, but they can be handy. For example, you can use one color for component designations and another for part numbers.
- How fast does the software perform tasks like component placement, screen redrawing, and printing? Of course, this is also a function of your hardware. Upgrading, recon- figuring, or replacing your computer might speed up slow software.
- Is the software copy protected? Does it require a hardware key? As a form of copy protection, some schematic draw- ing programs (mostly the higher-priced ones) require a special security key, sometimes called a *dongle,* to be installed on a parallel port connector in order for the soft- ware to run as shown in Fig. 2-27. If you don't want to bother with this, don't buy software that requires it.
- Does the software use on-screen menus, require that you type commands, or allow for both? On-screen menus are handy because it can be a chore to remember or look up the many commands used by a program. But typed com- mands can be quicker for the functions you use often. Many programs offer both.
- Does the software support mice, trackballs, or other input devices? A mouse is the most popular alternative to the keyboard, and can be very handy for use with schematic drawing, pc board layout, and other drawing software. A mouse allows you to easily move a cursor to a desired screen location, and mouse buttons enable you to quickly select items or perform other common functions.

Fig. 2-27 *Some software requires a security key, or "dongle," installed on a parallel port.*

- How good is the documentation and product support? Great features are useless if you don't know they're there or can't figure out how to use them. Product documentation includes user and reference manuals and on-screen help. Other product support might include a telephone help line, or an on-line bulletin board system (BBS) where you can post questions, look for tips, or download upgrades, component libraries, and other software.

Outputs

Eventually you'll want to print your schematic on paper or produce other output files for use in pc board design and fabrication.

- Which printers and plotters does the software support?
- Is the printed artwork scalable to different sizes? Are the sizes user-definable or fixed?
- What netlist formats are supported? If you want to import your schematic into a pc board design program, be sure your schematic capture software's netlist format is readable by your pc board design software.
- Are any other special file formats supported, in case you need to import the schematic to another program? Popular file formats include AutoCAD's DXF and Postscript.

Using general-purpose CAD for schematic drawing

Unlike dedicated schematic capture software, general-purpose CAD can be used to create many different types of precise, accurate drawings, including architectural drawings, drawings of mechanical parts and assemblies, graphs, woodworking or sewing patterns, and even works of art, as well as schematic diagrams and pc board artwork. General-purpose CAD enables you to create drawings with precisely placed lines and shapes, save the drawings to disk, and print or plot them on paper.

Almost all CAD software allows you to create and store symbols (also called objects or components). Instead of drawing a NAND gate from scratch each time, you can draw one NAND gate and store it on disk as a symbol, or component. The next time you want to place a NAND gate in a schematic, all you have to do is call up the symbol and place it at the desired position in your drawing. Some CAD programs offer symbol libraries of schematic and other symbols that are ready for use in your drawings, saving you the trouble of creating them yourself.

Like schematic capture software, general-purpose CAD ranges from basic, low-cost products to full-featured professional ones. Their main advantage is that they don't limit you to drawing schematics and pc board artwork. If you need CAD software for many purposes, and if you only occasionally draw schematics and pc board artwork, a general-purpose CAD program might fit your needs.

The versatility of general-purpose CAD is its main drawback when used for schematics and pc board artwork. Because the software has so many uses, it generally doesn't have features unique to schematic and pc board design, including design checking and netlist generation. And the component libraries usually aren't as extensive as those you'll find in dedicated schematic capture software.

At a minimum, general-purpose CAD software for drawing schematic diagrams should allow you to do the following:

- Place lines and curves at precise locations in a drawing.
- Create, save, and place symbols for resistors, capacitors, ICs, and other components in your drawings as needed.
- Print or plot drawings on dot matrix, laser, or other printers or plotters.
- Add text anywhere in the drawing.

Appendix A lists sources for general-purpose CAD software. Computer magazines occasionally review CAD software, and many have ads for new products, so these are a good place to look for information about the latest offerings.

Step-by-step: Drawing a schematic with Generic CADD

Generic CADD from Autodesk Retail Products is a moderately priced, general-purpose CAD program that can be used to draw schematics and pc board artwork. The following paragraphs describe Generic CADD as a schematic drawing program. Chapter 5 describes Generic CADD as a pc board layout tool.

Figure 2-28 shows Generic CADD in use for schematic drawing. Most of the screen is taken up by the drawing area, where the schematic is drawn. The drawing grid is defined by the network of tiny dots that covers the drawing area.

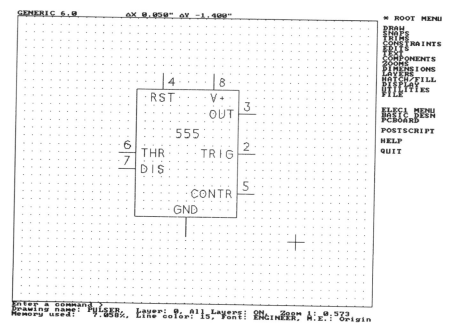

Fig. 2-28   *Generic CADD in use for schematic drawing.*

You can define your drawing area to be any size and zoom in or out to view the drawing, or any portion of it, at any scale. For schematics, the drawing area is usually defined as a standard paper size, such as 8½ by 11 inches.

Along the right side of the screen is the menu area, from which you can select functions. The main, or *Root,* menu is a list of submenus, with each submenu consisting of a list of related functions or commands. The mouse is used to select the menu items. These are the major submenus and their functions:

- *Draw.* For drawing lines, polygons, circles, arcs, and curves.
- *Edits.* For moving, copying, mirroring, or erasing a drawing or any portion of a drawing.
- *Snaps.* Causes the cursor to jump to the nearest point, intersection, or other selected location.
- *Trims.* For extending or trimming line lengths, and filleting and chamfering corners.
- *Constraints.* For turning the visible grid on or off, setting the grid size, forcing the cursor to snap only to grid points, and other control functions.
- *Text.* For selecting a text font, size, and spacing.
- *Components.* For loading, creating, and placing compo-

nents. *Component replace* is a handy feature that exchanges all instances of a particular component in a design with another selected component.

- *Zooms.* Allows you to view any portion of the drawing.
- *Dimensions.* For labeling dimensions. Mostly used for drawings of physical objects.
- *Layers.* For selecting, displaying, or hiding drawing layers. Each drawing can consist of up to 255 layers, one atop the other. Schematics require just one layer.
- *Hatch/Fill.* For creating solid objects or areas or filling an area with a pattern or hatch. Several hatch patterns are included or you can create your own.
- *Display.* For setting up the screen and selecting line types. You can select dotted or solid lines, line color, and drawing units (inches, millimeters). Cursor position can be displayed as absolute coordinates measured from the drawing origin or as delta coordinates measured from the previous cursor position.
- *File.* For loading, saving, printing, and plotting a drawing.
- *Menus.* For loading specialized menus.

Also included are conversion utilities for HP-GL plotter and DXF AutoCAD formats, and a bill of materials utility.

Generic CADD comes with a basic set of components for schematic drawing, but you might want to buy or create an expanded component library of schematic drawing symbols.

In Generic CADD, instead of using the menus, you can enter two-letter commands in the command line under the drawing area. Under the command line is an area with setting and status information about the drawing. Above the drawing area is a coordinates display that gives the current cursor position.

If you don't like the way the included menus are set up, you can arrange them as you wish. If you buy or create additional component libraries, you can add menus for these. The menus are stored as text files and are easily modified with any text editor.

A mouse (or trackball, digitizer, or other pointing device) moves a cross-hair cursor to draw lines and place objects in the drawing area and select menu items. The left mouse button controls the drawing area cursor, and the right or center mouse button controls the menu cursor.

The following steps illustrate how to draw a schematic using Generic CADD. The procedure below is one that has worked for

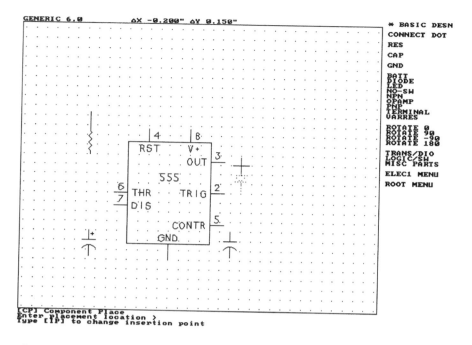

Fig. 2-29 *Placing components in a drawing.*

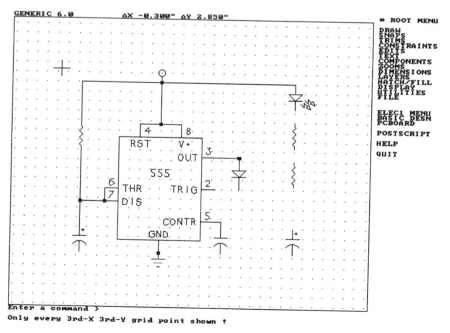

Fig. 2-30 *Adding connections.*

me, but you might have different preferences about grid size, order of placement, and other details. Other CAD software performs the same basic functions, but will vary in specific capabilities and other details.

This procedure assumes that Generic CADD is installed and configured for your computer, and that you are familiar with basic computer operation. It also assumes that a component library of schematic symbols (available from Generic CADD) has been installed and added to the menus.

1. Draw a rough sketch of your circuit on paper, or if you prefer to draw directly on screen, gather the data sheets or other information you will need to draw the schematic.
2. Run Generic CADD. The main drawing screen will appear.
3. Set the grid size to 0.1 inch or your preferred size. Set *Snap To Grid* on. This ensures that all lines and components are placed on grid points, so you can easily and quickly draw interconnections that meet precisely.
4. Set the drawing limits to 8½ inches wide and 11 inches tall. Zoom in to view an area of about 4 by 5 inches. The drawing limits are used by the *Zoom Limits* command to show you the entire drawing, but you can draw outside the limits, and you can change the limits whenever you want.
5. From the schematic components menu, select components and use the cursor to place them in the drawing (Fig. 2-29).
6. Draw the connections between the components (Fig. 2-30).
7. Number and label the components (Fig. 2-31).
8. Label the drawing. Include a title, date, your name, and any other identifying information you want on the drawing.
9. Save the drawing to disk.
10. Print the drawing (Fig. 2-32). With Generic CADD, you can print the entire drawing or any part of it. You can print at actual scale or any other scale that fits your paper. Dozens of dot matrix and laser printers are supported, or you can plot your schematic with a pen plotter.

Finding the schematic drawing software for you

Software products change rapidly. Falling prices and continued improvements in computer hardware mean that better compu-

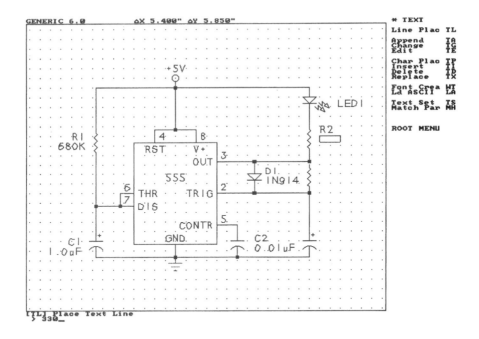

Fig. 2-31 *Labeling components.*

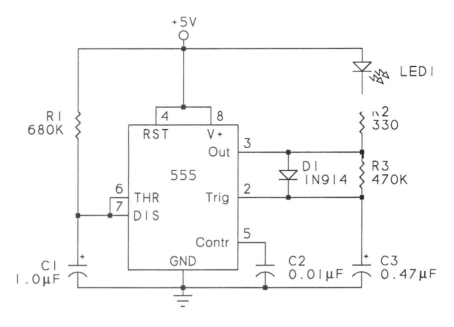

Fig. 2-32 *A printout of a schematic drawn with Generic CADD.*

ters and software are available at lower prices. No matter what type of personal computer you have, you can find a schematic drawing program to fit it and your budget.

If you're in the market for a schematic drawing program, check around to see what's currently available in your price range. In addition to the sources listed in appendix A, look for ads and articles in computer, engineering, and electronics magazines. Many companies offer demo disks for free or for a nominal cost, so you can get a feel for what the program can do before you buy. Some vendors offer a 30-day, money-back guarantee, so you can try the real thing before committing yourself.

If you're on a tight budget, investigate shareware and public domain programs. These are available on many on-line bulletin boards and from mail-order companies that specialize in shareware.

3
Component
choices

When you're ready to move from theory to reality, from a circuit drawn on paper to one built of real, three-dimensional components, you need to thoroughly define and describe the components that your circuit contains.

This book focuses on how to design and fabricate pc boards, rather than on the techniques of circuit design. For the most part, I assume that your circuit design is complete and that your circuit is ready to build. However, this chapter does cover some circuit design choices that relate directly to pc-board design, including choices in component packages and decisions about whether to mount components on the pc board or directly in the project enclosure.

Component packages

A component's package determines the physical dimensions of the component and the configuration of its leads or terminations. A single component might be available in several package types, including through-hole or surface-mount, plastic or ceramic, and high-power or low-power forms. Even a simple resistor can be ordered in a surface-mount or through-hole package and in a variety of power ratings and tolerances. Because the package defines the shape of a component and the spacing of its leads, you need to define the component packages before you can design the pc board artwork.

Which packages you choose depends on the electrical, thermal, and mechanical requirements of your circuit. Sometimes the choice is critical—when you need to fit your circuit into the tiniest space possible, or when you need a precise resistor value for an amplifier, for example. For a noncritical, one-of-a-kind circuit, you might get by with the components you have on hand, or you can use price as a guide to selecting components.

Component manufacturers and many distributors can supply package dimensions and other specifications for the components they offer. Some distributors' catalogs include enough information to enable you to make a choice. In fact, a good catalog from a distributor or mail-order vendor can go a long way in helping you learn what's available.

Manufacturers of ICs publish data sheets containing specifications for each IC they offer. Sometimes application notes are available with detailed design information, circuit ideas, and occasionally even pc board artwork or layout tips. Manufacturers of other components publish useful data about their products as well. Data is normally available on request and at no charge.

To design a pc board for a particular circuit, you should know, at least in a general sense, the length, width, height, and lead spacing for each component. This isn't as difficult as it might seem, because many components use standard packages. For example, all 14-pin DIP IC packages have pins in two parallel rows spaced 0.3 inch apart, with the pins in each row spaced 0.1 inch apart.

The following sections describe what to consider when choosing IC and semiconductor packages, as well as resistors, capacitors, connectors, and other circuit components.

Through-hole versus surface-mount components

Currently the two major types of component packages are through-hole and surface-mount technology (SMT). Many components are available in both forms.

For many years, through-hole was the leading package type for components on pc boards. Through-hole packages have leads, which are solderable legs or pins, that insert into holes on a pc board. On the bottom of the pc board, each lead solders to the pad that surrounds each hole. Resistors, capacitors, ICs, and just about all other pc board-mounted components are available in through-hole packages.

In many products, through-hole ICs are now being replaced by their equivalents in surface-mount packages. In fact, some newer components are available only as surface-mount devices (SMDs). Instead of having leads that insert into holes on the pc board, the surface-mount IC rests entirely on one surface of the board. Leads or terminations (leadless, solderable connections) are soldered to pads on the same side of the board where the component body rests.

Advantages of surface-mount components

Compared to through-hole components, surface-mount components have several advantages.

They save space. Surface-mount ICs have about half the package area of their through-hole equivalents. Figure 3-1 shows the same IC in a through-hole and surface-mount package. The surface-mount IC requires less than half the board area of the through-hole IC, and this size difference holds true for other surface-mount packages as well.

A board built with surface-mount components is typically 30 to 50 percent smaller than an equivalent board built with through-hole components. Smaller boards mean lower costs for the board and its enclosure.

Surface-mount components weigh less. A surface-mount IC might weigh $1/10$ or less of its DIP equivalent.

You don't have to drill holes. Because a surface-mount component rests on one surface of the pc board, there are no lead holes to drill. The only holes you need are for mounting hardware and vias (interlayer connections) on double-sided boards.

The smaller surface-mount packages have shorter propagation delays and lower parasitic inductance and capacitance. These can be critical in high-speed circuits.

Fig. 3-1 *Examples of a 14-lead IC in a through-hole DIP (top) and surface-mount SOIC (bottom) package.*

Disadvantages of surface-mount components

Surface-mount components aren't always the solution, however. They have several drawbacks.

The closely spaced leads and tiny pads of surface-mount devices leave little margin for error in hand-soldering. Surface mounting is well suited to automation, but hand-soldering requires precision and care. For example, the leads of a surface-mount small-outline IC (SOIC) are spaced 0.05 inch apart, twice as close as the 0.10-inch spacing on a through-hole DIP. Sloppy soldering of a through-hole joint might still result in a functional circuit, while the same sloppiness using surface-mount components might result in solder bridging between pads.

Because a surface-mount component has no mechanical connection to the board, it must be held in place with adhesive or another means while soldering. In contrast, through-hole components are held in place by the lead holes.

The narrower traces and closer spacing of surface-mount designs require greater precision in pc board fabrication. For a DIP IC with pads 0.04 inch wide, you can route a 0.02-inch wide trace between two pads and still have a 0.02-inch gap on each side. With surface-mount components, whose leads are spaced 0.05 inch apart or closer, the pads and spaces must be narrower, and it's much harder to squeeze a trace between two pads.

Traditional prototyping technologies like breadboarding and wire wrapping aren't suitable for surface mounting. It's hard to prototype a surface-mount design without designing and fabricating a pc board. One solution is to build a prototype with the equivalent through-hole parts. Adapter sockets are also available for breadboarding some surface-mount devices.

Surface-mount ICs are generally not designed for socketing, and so must be unsoldered for removal. This contrasts with DIP ICs, which you can mount in a low-cost socket for easy removal and reuse. Some surface-mount sockets are available, however.

Not all components are available in surface-mount packages. Some are available only as through-hole components. But this is changing, and eventually the through-hole versions will be less common.

Surface-mount components are more expensive than their through-hole equivalents. This too will change over time as surface-mount components become the norm.

Heat dissipation can be a problem in high-current, surface-mount circuits. Compared to through-hole circuits, surface-mount circuits run hotter for two reasons: the smaller surface-mount packages do not dissipate heat as effectively as their larger through-hole equivalents, and the closer spacing of surface-mount components increases the amount of heat generated per square inch of board space.

Removing soldered surface-mount components requires new tools and techniques. On a through-hole component, you can unsolder and free the leads one at a time, and then pry the component from the board. Instead of leads, some surface-mount components have rigid, solderable terminations that are difficult to completely desolder. On other surface-mount components, the leads are tucked under the package, making them difficult to access for soldering or desoldering. To remove a surface-mount device without destroying it, you often have to melt all of the connections at the same time, which requires special soldering iron adapters or a hot-air tool.

Making a choice

Whether you choose surface-mount or through-hole components depends on your circuit's requirements. Are you designing a lightweight controller for a radio-controlled glider? Then surface-mount is the natural choice. Are low component cost and ease of soldering critical? Then through-hole components make more sense.

Most designs fall somewhere in the middle. Often you can use either package type with no significant effect on circuit performance. Some designs use a combination of surface-mount and through-hole components.

The following sections describe popular package types for ICs, transistors, resistors, capacitors, and other components.

IC packages

Integrated circuits are available in both through-hole and surface-mount packages. Figure 3-2 shows examples of IC packages. A popular through-hole package is the DIP, whose inputs and outputs are brought out to leads along two sides of the package.

An advantage to DIP ICs is that they are easily socketed for removal or replacement. Instead of soldering an expensive IC to a board, you can solder a socket that matches the pinout of the IC. The IC then plugs into the socket, making an electrical and mechanical connection without soldering. If necessary, the IC can be pried out with a removal tool or a flat-head screwdriver, as shown in Fig. 3-3.

Fig. 3-2   *ICs are available in many package types.*

Fig. 3-3   *Removing a DIP IC from a socket.*

In addition to DIPs, other through-hole IC packages include pin grid array (PGA) packages, which are usually reserved for ICs with more than 40 pins, as well as the three-terminal packages described in the next section.

Surface-mount ICs are available in gull-wing, J-lead, and leadless packages. In a gull-wing package, the leads bend down and away from the body of the IC. This makes the leads easy to solder, inspect, and monitor with a test probe. A disadvantage is that the pc board must reserve room for the outward-bent leads and their solder pads. Specific gull-wing packages include the SOIC, with leads along two sides like a DIP, and the quad flat-pack (QFP), with leads along all four sides.

In a J-lead package, the leads are tucked under the package instead of bending out away from it. This conserves board space but makes the connections difficult to solder and inspect. The plastic leaded chip carrier (PLCC) is an example of a J-lead package with leads on four sides. Both through-hole and surface-mount sockets are available for PLCC devices.

Another package type is the leadless ceramic chip carrier (LCCC), which is shaped much like the PLCC, but instead of J-leads, it has metallized contacts in castellations or grooves along the package edges. The LCCC is reserved for high-reliability components.

Transistor packages

Transistors and other three-terminal devices are available in through-hole and surface-mount packages, as shown in Fig. 3-4. Popular through-hole packages include the TO-92, TO-220, and TO-3 packages. Each is designed for a different level of power dissipation. TO-220s mount horizontally or vertically on a pc board, as shown in Figs. 3-5 and 3-6. Horizontal mounting gives better support to the package, but vertical mounting saves board space.

Surface-mount transistor packages, or small-outline transistors (SOTs), vary in shape depending on the manufacturer and

Fig. 3-4 *Package types for transistors.*

Fig. 3-5   *A horizontally mounted TO-220 package.*

Fig. 3-6   *A vertically mounted TO-220 package.*

the power rating of the device. Two popular types are the SOT-23, for power dissipation up to 200 milliwatts, and the SOT-89, for dissipation up to 500 milliwatts.

Resistor choices

You can choose between through-hole and surface-mount resistors as well. Figure 3-7 shows examples of each.

With a through-hole resistor, the leads are bent to fit the holes on the pc board. Because the leads are flexible, precise hole spacing isn't critical. To fit a tight space, you can mount through-hole resistors vertically, as shown in Fig. 3-8.

A typical surface-mount resistor has a chip of ceramic for its base, with a resistive element screened onto it. In place of leads, chip resistors have solderable metal terminations at each end of the device. The terminations normally wrap around the package from top to bottom, allowing the resistor to be mounted either side up on the pc board.

Specific two-terminal surface-mount package types are designated by number, with the first two digits giving the length and the second two digits giving the width, each in hundredths of an inch. For example, the popular 1206 chip resistor is 0.126 inch long and 0.063 inch wide.

In addition to resistor package type, you need to be sure your resistors' power ratings and tolerances are adequate for your

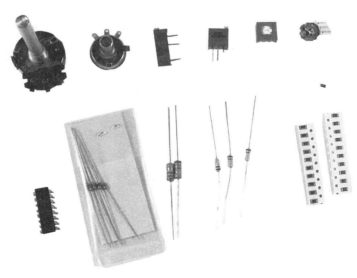

Fig. 3-7 *Through-hole and surface-mount resistors and potentiometers.*

Fig. 3-8   *To save board space, you can mount through-hole resistors vertically.*

application. Many applications are low power. For example, you can use ¼-watt or ⅛-watt resistors as pull-up resistors in logic circuits. In higher-power circuits, you need to choose resistors with adequate power ratings. The power a resistor dissipates equals the voltage across it multiplied by the current through it. Leave a generous margin of safety when choosing component ratings. For example, if you expect a resistor to dissipate 0.24 watt, use a ½-watt package rather than a ¼-watt package.

Resistors are available in wattages of ⅛, ¼, ½, 1, 2, and higher. Figure 3-9 shows examples of high-wattage resistors. Some high-wattage resistors are designed for chassis rather than on-board mounting. Chip resistors are limited to lower wattages.

Tolerance, or how close a resistor is to its rated value, is another consideration. Again, for many applications, tolerance isn't critical. Most modern resistors have tolerances of 5 percent, which means that the actual value should vary no more than 5 percent from the rated value. For example, a 1000-ohm, 5 percent tolerance resistor might in reality be anywhere from 950 to 1050 ohms. For critical applications, 1 percent and 0.5 percent tolerances are available.

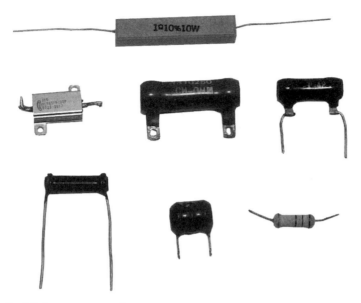

Fig. 3-9   *High-wattage resistors.*

Capacitor choices

As with resistors, capacitors offer a choice between surface-mount and through-hole packages. Figure 3-10 shows examples.

Cylindrical through-hole capacitors are available in two configurations: axial lead and radial lead. An axial-lead capacitor has a lead extending from each end and mounts on a pc board much like a through-hole resistor. A radial-lead capacitor has both leads extending from one end of the package. This saves board space, but requires more room vertically. A radial-lead capacitor is usually less expensive than the equivalent axial-lead package. Figure 3-11 shows an example of each.

Surface-mount capacitors are also available, in both chip and tubular packages. For large values, you might have to use a through-hole component.

Like resistors, capacitors have tolerances, as well as voltage ratings. Again, for some applications these values are more critical than for others. For example, in a critical timing circuit, you need a precise capacitor value, while precise tolerance for an input filter capacitor on a power supply is less critical.

Capacitors are also rated in working volts dc (WVDC), which is the dc voltage that can be applied across the capacitor for long periods without component damage. As a safety margin, choose WVDC values that are 50 percent or more above the highest voltage you expect across the component.

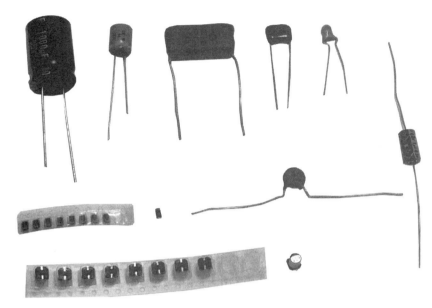

Fig. 3-10   *Through-hole and surface-mount capacitors.*

Fig. 3-11   *A radial-lead (front left) and axial-lead (front right) capacitor mounted on a pc board.*

Capacitors are constructed of many different dielectric materials, each with advantages and disadvantages. Briefly, here are some of the major capacitor types and their recommended uses

- Aluminum electrolytic—power supply filtering.
- Tantalum electrolytic—power supply decoupling, timing. Through-hole versions are sometimes called "teardrops" because of their shape.
- Polyester film, polystyrene, polypropylene—power supply decoupling, timing, filtering.
- Ceramic disc—power supply decoupling, timing, filtering.

Other components

Other components offer packaging options. Diodes, crystals, switches, relays, and potentiometers are available in a variety of packages, including surface-mount and through-hole. With these, as with all components, you need to consider the requirements of your circuit, what's available on the market, and your budget when deciding which specific components and package types to use.

Heat sinks

If your circuit contains components that will dissipate large amounts of power, you need to plan for heat-sinking these components. Heat sinks are devices made of metal or some other heat conductor that clamp or screw onto a component's case, drawing heat from the component, and preventing it from overheating. Heat sinks are available in many sizes to fit all standard component packages. If heat sinks are needed, you'll need to allow room for them on your circuit board, as shown in Fig. 3-12.

Components likely to require heat-sinking include voltage regulators and power transistors. The data sheets for your components should have information on power dissipation and heat-sinking requirements.

Component support

Many electronic components are lightweight and can be attached to the board by their solder joints alone. Heavier components might require brackets, clamps, or other supporting hardware that mounts on the pc board. If these are needed, be sure to include them in your pc board planning.

Fig. 3-12 *A TO-220 package with heat sink.*

On-board and off-board components

Some components don't mount on the pc board at all, but instead mount directly on the enclosure's floor or on the front or back panel. Common choices for off-board mounting include bulky items like transformers, as well as items that need to be accessible to the user, such as displays, switches, controls, power plugs, fuses, jacks, plugs, and connectors.

Before you draw your pc board artwork, you should decide which components will mount off-board and how they will connect to the pc board. On-board components require pads that match the configuration of the component package, while off-board components require pads to match the wire, cable, or connectors used.

Large components

Large, heavy components, such as transformers, can be mounted on or off the circuit board. With on-board mounting, you don't have to run wires from the component to the board. But the trade-off, of course, is the need for a larger pc board and possibly additional mounting supports for the board.

Indicators, controls, and connectors

In the final project, most pc boards are mounted inside a plastic or metal enclosure, with indicators, controls, and connectors mounted on the front and back panels of the enclosure. Consider carefully what you want to mount on the enclosure for easy access or viewing. An obvious choice for front-panel mounting is an on-off switch, along with any adjustments or switches you want available to the user. Don't forget to include displays— lamps, light-emitting diodes (LEDs), liquid crystal displays (LCDs), or other indicators, as needed. Connectors include power connectors, as well as inputs and outputs that connect to other circuits or devices. For example, the front panel of an audio amplifier might include an on-off switch and an LED to indicate power-on, a volume control, a level indicator, and input and output jacks. The back panel might hold a socket for an ac power cord.

Fuses

Fuses can be mounted on the circuit board or back panel, or you can use an in-line fuse holder. For easiest access, use a fuse holder that is accessible from the back panel.

Vibrating components

Use off-board mounting for any components that might cause the pc board to vibrate, such as fans or some relays.

Connecting to the pc board

Most circuit boards have some type of off-board wiring, whether it's to an ac line cord, displays, controls, or other circuits. Before you can draw your complete artwork, you need to decide what type of connections your circuit will use. Figures 3-13 and 3-14 show examples of pc boards mounted in enclosures along with their off-board connections.

One option for off-board wiring is to solder the connecting wires directly to pads on the pc board, as shown in Fig. 3-15. This method is low-cost because there are no connectors to buy, but you have to desolder or cut the wires if you want to break the connections—to remove a board for testing, for example. For easier disconnecting, use plug-in or snap-together connectors. The connector mounts on the pc board, or you can use an in-line connector with the wires soldered to the pc board.

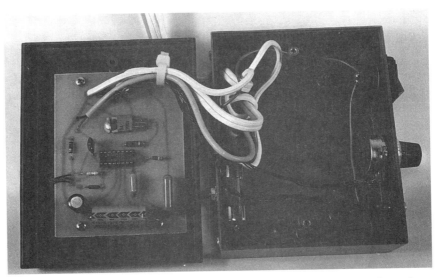

Fig. 3-13 *A pc board and wiring to front- and back-panel controls and connectors.*

Fig. 3-14 *A pc board and enclosure-mounted transformer.*

Some connectors use solder to attach the wires, while others use mechanical crimping. A ribbon cable is a good solution when you have a large number of low-power signals (Fig. 3-16).

Fig. 3-15 *Wires soldered directly to a pc board.*

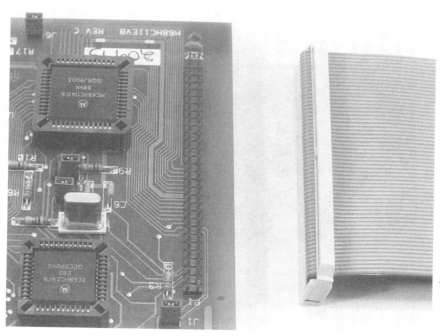

Fig. 3-16 *A ribbon cable and its pc board connector.*

Fig. 3-17 *A four-wire snap-together connector.*

Snap-together connectors with individually crimped wires are often used for power supply connections (Fig. 3-17). If you use ribbon cable connectors or other multipin, on-board connectors, you'll need to plan your pc board layout so that all of the off-board connections come together at the connector's pads on the pc board.

Sometimes it's possible to design the enclosure and the pc board to work together as a unit. With this type of design, the controls, indicators, and external connectors mount on the main pc board and pop into matching holes on the enclosure. This approach saves the trouble of wiring the controls and indicators separately, but you must plan the pc board layout carefully so that everything ends up in its desired location. Figure 3-18 shows an example.

Testing and verifying your design

A design is much easier to change on paper than after you've designed and fabricated the pc board. Often you can save time and trouble by building a prototype, or working model, of your circuit, so you can test it before you design the artwork and make the pc board.

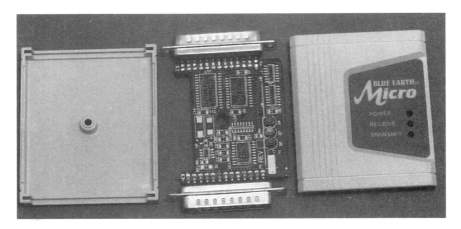

Fig. 3-18   *On this pc board, the connectors and three LEDs are designed to pop into matching openings on the enclosure, eliminating the need for cabling.*

For many circuits, breadboarding, wire wrapping, or point-to-point soldering will give you a good idea of how the circuit will perform when it is built on a pc board. The difference in construction methods will have no noticeable effect on circuit performance.

For simple circuits, the quickest prototyping method is solderless breadboarding. This method also allows you to pull the components and wires off the breadboard to modify your design, try different component values, or reuse the components when you're finished with the prototype. For a complex circuit, you might find it easier to wire wrap or point-to-point solder the circuit.

You might find it just as easy to go immediately to a pc board for the prototype, repairing and modifying the layout as necessary, rather than laboriously building a prototype using a different method. High-frequency circuits, including radio or microwave frequency circuits and high-speed digital circuits, must be built on a pc board.

If you do build a prototype, be sure to test it thoroughly at all of its expected extremes of input voltages and currents, temperatures, and so on. Vary the inputs through their entire desired range of operation and monitor the outputs. If the circuit must operate over a range of temperatures, test its performance at those temperatures. Monitor critical signals for noise. Verify that the heat sinks are adequate. When you're confident that your circuit design is complete, you're ready to begin the artwork for the pc board.

4

Techniques for designing pc board artwork

When you are confident that your schematic is complete and correct, and you have a complete parts list to go along with it, you're ready to design the pc board artwork. The artwork is an accurately scaled image of the copper pattern that will be etched onto the pc board. The artwork defines the locations of the traces, pads, and holes on the board. After the artwork is designed, you can use any of a variety of techniques to transfer the image to the pc board for etching.

As with schematic drawing, you can draw pc board artwork by hand or with the aid of a computer. But no matter which method you choose, laying out the artwork requires patience and attention to detail. Usually, some trial and error is involved as you experiment with different arrangements of components and traces. Careful checking is essential to ensure that the artwork is complete and accurate. A single misplaced trace might result in a circuit that doesn't work, possibly damaging or destroying components.

This chapter introduces the basics of designing pc board artwork, whether by hand or with the aid of a computer. Topics include how to estimate board size, choose trace widths, determine pad and hole sizes, and route signal and power traces. Following this, chapter 5 describes a variety of methods you can use to create the artwork.

Before you begin

Spend some time planning your artwork before you begin to draw it. You can save yourself time and effort in the long run.

Before you begin to draw, you need to estimate your board's size, trace widths, and pad sizes and shapes.

Estimating board size

A preliminary estimate of your pc board's size will help you as you place component pads in the artwork. Some circuits must fit into a predefined board size and shape. For example, a computer expansion board that plugs into a socket on a motherboard must fit the existing socket and enclosure.

For many projects, there is no predefined board size, but an estimated size will give you an idea of how the final project will look. If necessary, you can often condense a preliminary layout onto a smaller board in the final design.

One simple, effective way to estimate board size is to gather up the actual components that will mount on the board, including ICs or IC sockets, heat sinks, connectors, resistors, capacitors, and connectors. Arrange the components on a flat surface, preferably a sheet of graph paper with a 0.1-inch grid, as shown in Fig. 4-1. For a noncritical, single-sided board, leave about 0.5 inch between ICs and 0.1 to 0.2 inch between resistors, capacitors, and other components. This will give you a rough idea of the board size you'll need. For double-sided boards or circuits with narrow traces, you can space the components more closely.

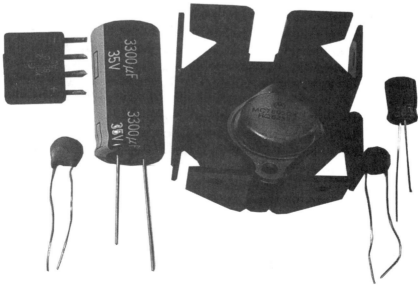

Fig. 4-1 *To get an idea of how big a board your circuit will require, arrange the components on a sheet of graph paper.*

If gathering the actual components is impractical, you can sketch the component packages on paper, arrange their shapes on a computer screen, or just estimate the board size from the number of components and their package sizes. You can even buy or cut your own paper or cardboard shapes to represent the components and arrange them on paper to give you an idea of how everything will fit.

Inspecting similar pc boards is another way to get a general idea of the board size you will need. Look for boards that have the same number of conductor layers as your planned board, similar trace widths to those you plan to use, and component types similar to yours.

If you need to install the circuit into a certain-size space or enclosure, estimating the board size will give you an idea of whether the circuit will fit. Be sure to consider the space above and below the board as well as the board's surface area. This includes under-board clearance for leads and above-board clearance for tall components, as Fig. 4-2 illustrates. You don't want to discover at the last minute that a radial-leaded capacitor is too tall to fit into the enclosure, or that a relay's package overlaps an adjacent component on the board.

Also remember to allow room for edge connectors, card guides, screws, and other hardware needed to mount the board in its enclosure. A board that mounts with screws and standoffs requires a hole in each corner. Larger boards might require additional mounting holes for support. If the board will slide into guides in a card cage, be sure to leave the board's edges free from components and solder joints so the board will slide into the guides without interference.

Often, it's convenient to plan the layout to fit a standard, precut board size. Common sizes are 3 by 6 inches, 4 by 6 inches, 6 by 6 inches, and larger. Check your suppliers' catalogs to see what sizes are available. If you need a nonstandard board size, try to plan so you can cut a larger board into multiples, without waste. For example, for a circuit requiring a 3-by-3-inch board, you can easily get two copies from a single 3-by-6-inch board, or use the second half of the board for another circuit. But if you cut two 3-by-3-inch boards from a 4-by-6-inch board, you'll be left with a less-useful 1-by-6-inch strip.

If your board will be larger than 8 by 10 inches, consider dividing it into two or more modules and connecting them with cables. The larger the board, the harder it is to transfer the artwork without distortion or other problems and the harder it is to etch the board evenly.

Fig. 4-2 *As you plan your layout, remember that cylindrical capacitors, heat sinks, relays, and other components might be wider than their pads would suggest. Also be sure to allow room in your enclosure for tall components.*

Double-sided boards

Before you draw your artwork, you need to decide how many artwork layers your board will have. On a double-sided pc board, both the top and bottom surfaces of the board hold artwork. In most double-sided boards, all of the components rest on one side of the board.

You might decide to use a double-sided board for a complex circuit, especially if it has to fit into a small space. But double-sided boards are more expensive and require special techniques for layout and fabrication.

Connections between layers on a double-sided board are made with *vias,* or feed-throughs, which are small, drilled holes that terminate in small pads on both sides of the board. In commercially fabricated boards, the walls of the holes are plated with a conductive material that electrically connects the two pads. This type of via is called a *plated-through hole.*

For hand-fabrication, you can make the interlayer connections by inserting small lengths of wire into the vias and soldering them to the pads, or by installing eyelets, which are flanged, tubular metal conductors designed to simulate plated-through vias. Figure 4-3 illustrates the options. Component lead holes can also act as vias.

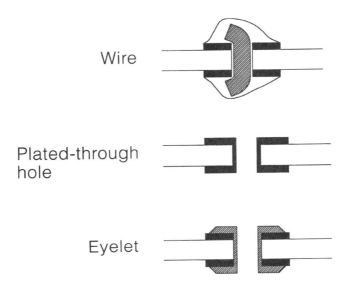

Wire

Plated-through
hole

Eyelet

Fig. 4-3 *To connect traces on the top and bottom layers, you can use a*
wire or component lead inserted into a hole and soldered on
both sides, a plated-through hole, or an eyelet.

On a double-sided board, the two artwork layers must line
up accurately one above the other. A hole drilled in the center of
a pad on the top of the board must emerge in the center of its
corresponding pad on the bottom. When it does, the board is
said to have accurate *registration,* or side-to-side correspondence
between the layers. The more artwork layers there are, and the
smaller the pads, the harder it is to achieve accurate registration.

The artwork for a double-sided board should contain at least
three widely spaced alignment holes. These are used for lining
up the top and bottom artwork when the board is fabricated. You
can use vias, component through-holes, or mounting holes as
alignment holes, or you can add dedicated alignment holes that
have no other purpose.

You can make good-quality, double-sided boards without a
large investment in special equipment. But fabrication of boards
with more than two layers is best left to specialists. You can
design the artwork for these boards yourself, and send them out
for fabrication as described in chapter 7.

Arranging the components on the board

When you are ready to begin drawing the artwork, where do you
start? Before you route any connections, you must place the com-

ponents on the board. Where the components are located on the board and where they are in relation to each other can affect how the circuit performs, how easy it is to route the traces, and how convenient it is to test and repair the finished board.

Some component positioning requirements are unbreakable, while others are desirable, but not essential. For most circuits, there is no one correct layout; a variety of approaches will give good results. Often you'll have to make trade-offs—for example, between the simplest or most compact layout and the most convenient layout for assembly, testing, or use.

Some components and circuit elements, such as edge connectors, have fixed, inflexible locations on the board. Place these first in your layout. Other components might have preferred locations, and these should also be placed early in the layout process. For example, a potentiometer that users will need to access should be placed in an easily accessible location on the board, such as along the top edge of a vertically mounted board.

Sometimes component placement can affect circuit performance. For example, an op amp's decoupling capacitors should be as close as possible to the pins they connect to. Voltage regulators might require extra clearance for heat dissipation. Be sure to keep heat-sensitive components, such as timing crystals, away from heat-radiating components. Check component data sheets for notes on critical component placements for particular devices.

As in the schematic, try to group related circuit elements together on the artwork. Power supply components, digital logic blocks, and analog circuit blocks are examples of component groups that should be grouped in the layout. Doing so makes the circuit easier to lay out, test, and troubleshoot, and might also improve circuit performance.

Often the schematic contains clues about how to arrange the components on the board. Many circuits have an overall direction of signal flow, from inputs to outputs, and from left to right on the schematic. In an amplifier, for example, signal flow is from low-level input to amplified output. Often the artwork can follow the same overall flow as the schematic.

If two components share many connections, such as a microprocessor and memory ICs, these should be located near each other on the layout. In addition, try to orient all components along a grid of perpendicular lines, as shown in Fig. 4-4. Unless you have good reason not to, orient all components so the packages are parallel to the board edges, for neatness and ease in locating parts on the board.

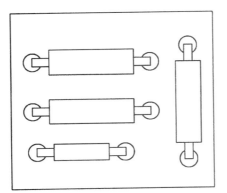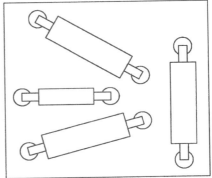

Fig. 4-4   *For neatness and ease of use, orient components so that their packages are parallel with the board's edges (left), not tilted at random angles (right).*

Also, as much as possible, position all ICs and polarized components so they are oriented in the same direction. For example, place all DIP ICs on the board so that pin 1 of each points toward the same corner of the board. This type of consistency will help prevent errors in circuit assembly and testing.

Trace widths and spacing

How wide and how far apart should your board's traces be? This depends on the amount of current the traces must carry. It is also limited by the abilities of your artwork design and board-fabrication techniques.

As a general rule, use the widest traces and widest spacing that you can, while still achieving the layout and board size you want. Boards with wide traces spaced far apart are easier to fabricate than boards with narrow traces spaced close together. When the traces are wide, a less-than-perfect etching job is less likely to result in an open trace or a trace that is too narrow for the required current. When the traces are widely spaced, you're less likely to find unwanted short circuits caused by incomplete etching or solder bridging between traces. Wide traces can also reduce electrical noise problems because of their lower impedance.

As with copper wire, the amount of current a pc board trace can safely carry is a function of its cross-sectional area. Thick, wide traces can carry more current than thin, narrow ones. Table 4-1 shows the amount of current that traces of varying widths and thicknesses can safely carry. The limits are determined by the temperature rise and voltage drop along the trace. As a trace

Table 4-1   Maximum currents for different trace widths.

Weight of copper foil (oz./ft^2)	Trace width (in.)	Maximum current (amperes)
0.5	0.005	0.13
	0.010	0.50
	0.020	0.70
	0.030	1.00
1	0.005	0.50
	0.010	0.80
	0.020	1.40
	0.030	1.90
2	0.005	0.70
	0.010	1.40
	0.020	2.20
	0.030	4.00

conducts current, its temperature increases and a voltage difference develops between one end of the trace and the other. The wider the trace, the less the temperature rise, and the smaller the voltage drop.

Many traces carry a few milliamperes at most, and for these you don't have to worry about traces being too narrow to carry the current. Even narrow, 0.01-inch wide traces on 1-ounce copper can safely carry 800 milliamperes.

Narrow, closely spaced traces do have some advantages. They require less room on the board, and they can be routed between IC pads more easily. For example, a trace that runs between two adjacent pads on a DIP IC must be about 0.02 inch or less in width.

Spacing between traces becomes more critical when there are large voltage differences between traces. Table 4-2 shows

Table 4-2   Minimum trace spacings (uncoated, 0–10,000 ft. altitude).

Voltage difference between traces (dc or ac peak volts)	Minimum spacing between traces
0–50	0.015 in.
51–150	0.035 in.
151–300	0.050 in.
301–500	0.100 in.
500+	0.0002 in./volt

OK

Not recommended

Fig. 4-5 *For good solder flow onto the pad, traces should be narrower than the pads they connect to.*

recommended spacings for traces according to their difference in peak voltages. Low-voltage signal traces can be spaced 0.015 inch apart, while higher-voltage traces require wider spacing. Coated boards can tolerate closer spacings.

For good solder flow onto the pad, traces should be narrower than the pads they connect to, as shown in Fig. 4-5. With oblong or oval pads, the traces should, when possible, contact the longer side of the pad, but this isn't always feasible.

Minimum trace widths and spacings are limited by the techniques you use to design and fabricate your board. For example, if you use direct-transfer tapes to create the board layout, you're limited to the tape widths available. Some CAD software allows infinite choices of trace width. In this case, the limits of your fabrication techniques will define how thin your traces can be.

Generally, trace widths of 0.02 inch with 0.02-inch spacing are feasible with any fabrication technique. Narrower traces and closer spacing are possible, but you'll want to experiment and be sure that your fabrication methods can achieve good results with sizes or spacings narrower than this.

As a general guideline, for a typical board with through-hole components, you can use 0.04-inch traces for high-current power and ground traces, 0.02-inch traces for low-current signals, and 0.02-inch spacing between traces. A circuit that must fit in limited space, or uses surface-mount components, or requires traces between IC pads, might need narrower traces or closer spacing. And high-power, high-voltage circuits require wide and widely spaced traces.

Other guidelines for routing traces include

- When two or more traces run between rows of pads, space them evenly in relation to each other and the pads.

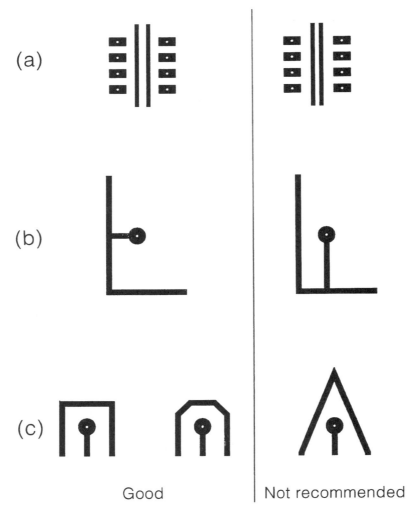

Good | Not recommended

Fig. 4-6   *Routing guidelines: (a) Space traces evenly in relation to each other and to pads. (b) Use the shortest routings possible. (c) Avoid angles of less than 90 degrees.*

Fig. 4-7   *Corners can be rounded (filleted), angled (chamfered), or square.*

- Use the shortest routings possible between pads.
- Route traces only at angles of 90 degrees or greater (see Fig. 4-6).

Corners

If you examine a variety of pc boards, you'll see that on some, the traces make sharp, 90-degree corners, while others use angled (chamfered) corners, and still others use rounded (filleted) corners. Figure 4-7 shows examples of each.

Rounded or angled corners conserve board space and are less likely to form solder bridges with adjacent traces or pads. In addition, the copper foil is less likely to crack or lift from a rounded trace. In high-frequency circuits, rounded corners can prevent the reflections and radiation losses that can occur with square corners. However, square corners might be easier to draw, especially when using transfer tapes, and they perform adequately in most circuits. In most cases, you can choose the type of corner that is most convenient.

Pad sizes and shapes

Pads, or lands, are the copper areas to which the component legs, leads, or terminations are soldered. Pads, like traces, can vary in size, shape, and spacing.

The Institute for Interconnecting and Packaging Electronic Circuits (IPC) is a nonprofit trade association that develops and publishes standards and specifications for pc board design and fabrication. It publishes recommended pad dimensions and spacings for standard component packages. You can also get ideas for pad sizes and shapes by examining the pads on existing boards and pc board artwork. If your transfer patterns or software contain predefined pad shapes for components, you can use these in your artwork.

Through-hole pads   The ideal pad size for a through-hole component depends on the size of its lead holes. The minimum recommended diameter for lead holes is the maximum lead diameter plus 0.006 to 0.020 inch. For example, a resistor with a lead diameter of 0.024 inch requires a lead hole with a diameter between 0.030 inch and 0.044 inch. Table 4-3 shows the diameters of different gauges of solid wire.

For ICs and any component where there are multiple, fixed-location leads, holes in the larger end of the allowable range are preferable. The extra room in a larger hole will help compensate for drilling errors and make it easier for all of the leads to slide into their holes at once. Leaving room for error is especially important for hand-drilled holes. For resistors and similar components, a large hole is less important because you can easily bend the leads to fit the holes' locations.

Table 4-3   Solid wire diameters.

AWG (American Wire Gauge)	Diameter (inches)	Maximum current (amperes)
10	0.1019	14.8
11	0.0907	11.8
12	0.0808	9.3
13	0.0720	7.4
14	0.0641	5.8
15	0.0571	4.6
16	0.0508	3.6
17	0.0453	2.9
18	0.0403	2.3
19	0.0359	1.8
20	0.0320	1.4
21	0.0285	1.1
22	0.0253	0.91
23	0.0226	0.72
24	0.0201	0.57
25	0.0179	0.45
26	0.0159	0.36
27	0.0142	0.28
28	0.0126	0.22
29	0.0133	0.18
30	0.0100	0.14

From the lead hole's size you can calculate the pad's diameter. For easy fabrication and soldering, the width of the annular ring (the circle surrounding the drilled hole) should be at least 0.012 inch. Using this guideline, a lead hole with a 0.03-inch diameter requires a pad with a diameter of at least 0.054 inch. This is a minimum value. Wider pads are easier to fabricate and solder. On the other hand, wider pads are more difficult to route traces between, and if a pad is too large, the solder will tend to wick away from the lead onto the large surface of the pad. Figure 4-8 illustrates pad sizing for through-hole components.

If you don't want to bother measuring lead diameters, you can drill test holes with different bits until you find a size large enough for the leads to fit through. To calculate the pad diameter, add at least 0.024 inch to the hole diameter.

If you will be drilling your pc board by hand, include holes in the artwork pads if possible. An etched hole in the pad is a visual reference during drilling, and also helps to physically guide the drill bit into the center of the pad. The size of the holes in the artwork pads doesn't have to match the size of the drilled hole. In fact, a tiny, just-visible etched area in the center of the pad does the best job in centering the drill bit.

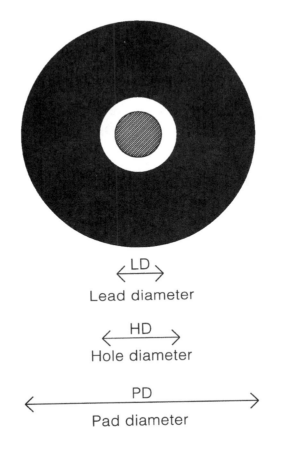

LD
Lead diameter

HD
Hole diameter

PD
Pad diameter

$$LDmax + 0.006 <= HD <= LDmax + 0.02$$

$$PD >= HDmax + 0.024$$

LD(in.) max.	HD(in.) max.	PD(in.) min.
0.018	0.024	0.048
0.024	0.030	0.054
0.032	0.038	0.062

Fig. 4-8 *How to calculate hole and pad size for through-hole components.*

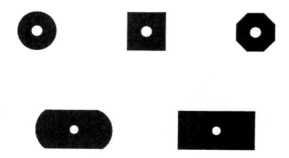

Fig. 4-9   *Component pads can have a variety of shapes.*

When pc boards are fabricated commercially, the holes are drilled by numerically controlled (N/C) machines that are programmed with the physical location of each hole on the board. If your board will be fabricated using such a machine, the component pads do not require holes.

Most pc board design software comes with predefined pad shapes. With some software, the through-hole pads include center holes; others do not, or allow a choice. Shapes for pads include round, square, rectangular, and oval (Fig. 4-9). Round pads are easily soldered because the pad's width is uniform all around the lead. Oval or oblong pads are used on DIP ICs to give a wide soldering surface on two sides while allowing clearance between adjacent pads within a row.

If you print your artwork on a low-resolution printer, the curved edges on tiny round or oval pads might appear jagged, and you might prefer to use square or rectangular pads for this reason. This is rarely a problem on laser printers with resolutions of 300 dots per inch or greater.

To indicate pin 1 on an IC, you can use a different pad shape or add a dot next to pin 1, as Fig. 4-10 illustrates. For many components, a range of pad sizes will work. Often you can design the artwork for an entire board using two or three pad and hole sizes. Keeping to a few sizes simplifies artwork creation, especially if you lay out the artwork by hand. Limiting the number of hole sizes saves time during drilling, with fewer changes of drill bits. In computer-aided artwork design, varying the pad sizes might cause little or no inconvenience, so limiting the number of variations is less important.

Surface-mount pads   Pads for surface-mount components don't require holes, but their sizes and shapes can affect how easily the component solders. Ideally surface-mount pads help to center the component on the pads during soldering. Pads that are

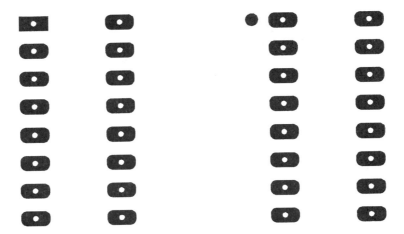

Fig. 4-10 *Two ways to indicate pin 1 on an IC, using a different-shaped pad (left) or a dot (right).*

too long tend to cause the component to float to one side while being soldered, while pads that are too wide tend to cause the component to rotate. Pads that are too short or thin are hard to solder at all.

The ideal pad size and shape for a surface-mount component varies slightly with the soldering method used, whether it's hand-soldering, wave soldering, or reflow soldering. But creating a different pad shape for each soldering method is inconvenient. As a compromise, the IPC has developed and published formulas for creating standard pad shapes for different component packages. The shapes are designed to give good performance, regardless of soldering method, for all components that fall within the specified size range.

Figures 4-11 and 4-12 show examples of recommended pads for popular surface-mount resistors, capacitors, and ICs. If you have direct-transfer patterns or predefined pads in the software you are using, you can of course use these.

Scaling the artwork

With many artwork design methods, you can create your artwork at larger-than-life scale. Printed circuit board design software often allows you to view and print the artwork at different scales. You can zoom in and view the artwork at an enlarged scale to place pads and lines precisely where you want them, then zoom out to view an entire layout at reduced scale. Usually you can print the artwork at actual or enlarged scales.

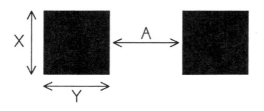

Resistor Designation	X (in .)	Y (in .)	A (in .)
RC 0805	0.055	0.060	0.032
RC 1206	0.063	0.063	0.070
RC 1210	0.102	0.063	0.070

Capacitor Designation	X (in .)	Y (in .)	A (in .)
CC 0805	0.055	0.060	0.032
CC 1206	0.063	0.063	0.070
CC 1210	0.102	0.070	0.070
CC 1812	0.126	0.070	0.126
CC 1825	0.260	0.070	0.126

Fig. 4-11 *Determining pad size and spacing for surface-mount resistors and capacitors.*

If you are using photographic methods, where the image can be easily reduced or enlarged, you can improve the quality of the image by drawing it to an enlarged scale. For example, artwork tapings done by hand typically have errors of up to 0.015 inch, because placing the pads and traces more precisely than this by hand is difficult. But if you tape the original artwork at ×2 scale, photograph it, and then print it at ×1 scale, errors of 0.015 inch are reduced to 0.0075 inch. Using ×4 scale gives even better results, down to 0.0038 inch error.

If you draw the artwork at ×2 scale, the result will have four times the area of the actual-size artwork. For example, a 4-by-5-inch board becomes 8-by-10 inches at ×2 scale, with an increase in area from 20 to 80 square inches. At ×4 scale, the artwork is 16-by-20 inches, or 320 square inches.

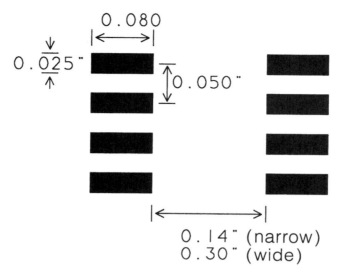

Fig. 4-12 *Determining pad size and spacing for SOIC components.*

Pad spacing

Just about all components have two or more leads or terminations and require multiple pads. For through-hole components, the leads must pass through the centers of the pads, and for surface-mount components, the leads or terminations rest on top of the pads. For a component to fit its pads properly, the pads must be spaced correctly on the board.

Most pc board design software includes patterns for correctly spaced pads for common components, and press-on and rub-on patterns include prespaced pads for many popular components. Even so, you might find you have to determine spacing for new or unusual components.

Pad spacing for some components is easy to determine. Dual in-line pin ICs typically have leads in rows spaced 0.3 or 0.6 inch apart, with the leads in each row spaced 0.1 inch apart. Other IC packages have predefined lead spacings, which are described in their data sheets.

For many radial-lead, through-hole components, including capacitors and transistors, catalogs from manufacturers or vendors might include lead spacings that you can use to define your pad spacings, or you can measure the lead spacing on the actual component.

The leads of axial-lead components, which extend horizontally from the component body, must be bent before inserting. To prevent damage to the component, the leads should extend horizontally before bending, and the pad spacing should allow for this. Figure 4-13 shows calculations for pad spacing for axial-lead components, including resistors and capacitors. For convenience in board design, round the lead spacing up to the next highest 0.05 or 0.01 inch, and bend the leads to match.

Ground planes

Some pc boards include a ground plane, which is a large area of copper that provides a low-impedance circuit ground. In multi-layer boards, an entire layer might be devoted to a ground plane. Copper areas might also be used as shielding for sensitive components. A side benefit of leaving large areas of copper on the board is conservation of etchant, because these areas remain unetched.

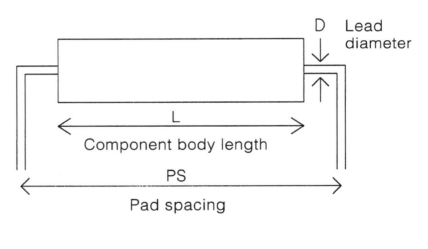

$$PSmin \geq Lmax + 7Dmax$$

L (in.) max.	D (in.) max.	PS (in.) min.
0.16	0.018	0.3
0.28	0.024	0.5
0.39	0.024	0.6

Fig. 4-13 *Determining pad spacing for axial-leaded, through-hole components.*

Designing the layout

When you've made your decisions about pad sizes and spacing, you're ready to draw the artwork. This involves placing components in the artwork and laying the traces to connect the component pads. The following sections describe methods for doing so.

Orientation of pc boards

Traditionally pc boards have been described as having a component side, on which the components are mounted, and a solder side, containing the pads to which the component leads solder. These descriptions are accurate for boards that use through-hole components, but they don't fit as well when describing surface-mount boards. On these boards, the solder is applied to the same side of the board that the components rest on, so the solder and component sides are one and the same.

In the interest of accuracy and to avoid confusion, in this book I'll refer to the two surfaces of a pc board as the top and bottom. On a single-sided board, the top is the traditional component side, which might have through-hole or surface-mount components mounted on it. The bottom is the traditional solder side, where through-hole components are soldered. On a single-sided, surface-mount board, the bottom is blank, except for possible jumper wires.

On a single-sided board that uses both surface-mount and through-hole components, the surface-mount components are soldered to the bottom. On a double-sided board, through-hole components are mounted on the top, and surface-mount components might appear on either or both sides. With multilayer boards, the layers are usually distinguished by number, with the top as layer 0 or 1.

Besides keeping track of which side of the board is which, you need to remain clear about which orientation of the artwork you are viewing. Designing and fabricating a pc board often involves working with mirror images of the artwork that will be etched onto the board.

On a single-sided, through-hole board, there is just one piece of artwork, which is etched onto the bottom of the board. When you look at the artwork on the bottom side, you see a mirror image of how the components are installed when viewed from the top side of the board. One way to think of it is that on the top, the pins of a DIP IC count counterclockwise around the chip, while on the bottom, they count clockwise, as shown in

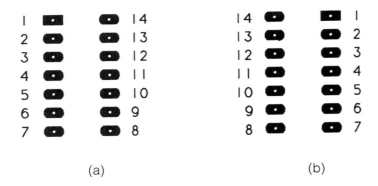

Fig. 4-14 *(a) In the top, or component-side, orientation, the leads on a DIP IC count up counterclockwise around the chip. (b) In the bottom, or solder-side, orientation, the leads count up clockwise.*

Fig. 4-14. On a single-sided, surface-mount board, the artwork is on the top, or component side, so the component orientation and artwork orientation are the same.

The artwork orientation you see on the board is sometimes called a right-reading image, while the mirror image is called a reverse-reading image. If the artwork contains any text or numbers, the orientation is easy to identify. On a right-reading image, any text in the artwork is oriented correctly, while on a reverse-reading image, the text is mirror-imaged. This is one good reason to include a board identifier, the designer's initials, or some small amount of text on every piece of artwork.

When you transfer the artwork to the pc board, you have to be sure it will end up in the desired orientation on the board. Most designers draw the pc board artwork as a top-side image, with the components oriented as you would see them on the board, even though the actual bottom-side artwork is a mirror image of this.

Different image transfer methods require artwork with different orientations. Several methods exist for mirror-imaging artwork from one orientation to another. Most of the time, you can draw your artwork in whatever orientation you're comfortable with (usually the top-side orientation), and mirror it later, if you need to.

Step-by-step:
The drawing process

The method you use to create your artwork, whether it's by hand with transfer tapes and patterns or with the aid of a computer, affects how you approach the drawing process. But some general guidelines hold true no matter what methods you use.

Except for the simplest circuits, pc board artwork is not something you draw in a single process, placing components and laying traces one by one until the board is finished. Instead, most artwork undergoes many revisions before it is complete. You might move a component or trace several times before placing it in its final location.

You can draw rough drafts of your layouts on paper or on a computer screen. Whether on paper or computer, different colors are useful for making power supply and ground traces stand out, and for distinguishing between board layers on a double-sided or multilayer board. For example, you can draw all top-side traces in red and all bottom-side traces in blue.

The exact procedure for creating the artwork varies, depending on your preferences and your circuit. In general, the drawing process goes something like this:

1. Place pads for the components with fixed or desired locations, if any (Fig. 4-15). These might include connectors, displays, and adjustable components that must be accessible.

2. Place pads for other components in their estimated locations on the board (Fig. 4-16). Using the schematic as a guide, try to group interconnecting components near each other in the artwork. You might want to place only the ICs and other major components now, and add resistors, capacitors, and other discrete components as you draw their traces. For large circuits, you might want to lay out the artwork in stages, placing components and drawing traces for small blocks of components at a time.

3. Draw power supply and ground traces to the components you've placed (Fig. 4-17). Rearrange the components as necessary to allow these traces to be drawn. Also draw any traces with special routing requirements.

4. Draw the short signal traces, adding and rearranging components as necessary (Fig. 4-18).

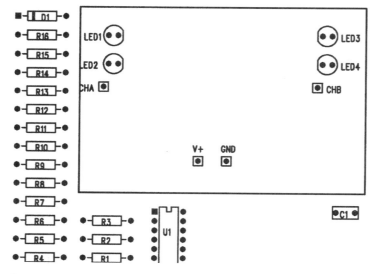

Fig. 4-15 *Place the pads for components with fixed or preferred locations on the board.*

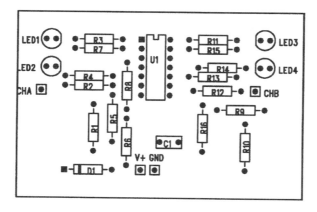

Fig. 4-16 *Place the pads for other components.*

5. Draw the long signal traces, adding and rearranging components as necessary (Fig. 4-19).
6. Place the next group of components, if any, and repeat steps 2 through 6.
7. Check and revise as necessary until the artwork is complete.

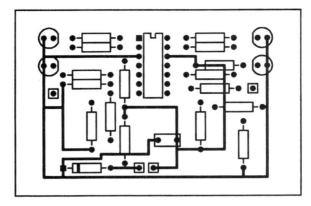

Fig. 4-17 *Route the power supply and ground traces, as well as other critical traces.*

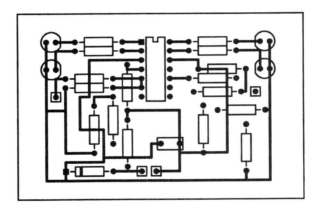

Fig. 4-18 *Add the short signal traces.*

Blocked traces

As you draw the traces in a layout, eventually you'll find yourself trying to route a trace that just can't get to where it has to go because other components or traces are blocking the way. Here are some ways to deal with the problem (Figures 4-20 and 4-21 illustrate):

- Examine the traces that interfere. Can they be rerouted so that the blocked trace can reach its destination?
- Move one or more components to allow the trace to be routed. Of course, this must be done carefully because it might create new routing problems. But if your original component placement was not ideal, moving things around might help.

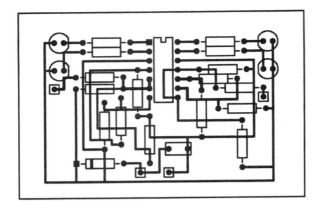

Fig. 4-19 *Add the long signal traces.*

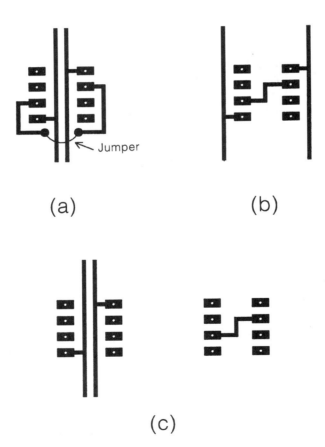

Fig. 4-20 *To route a blocked trace, you can (a) add a jumper, (b) move interfering traces or components, or (c) use a double-sided board.*

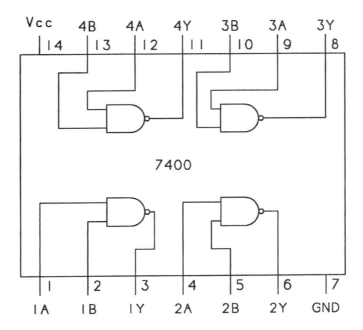

Fig. 4-21 *Many ICs, like this 7400 NAND gate, contain multiple, identical elements that you can swap to increase your routing options.*

- Swap pins or parts. For example, a 7400 chip contains four identical NAND gates, each with two identical inputs. Sometimes swapping gate or pin assignments will free up a circuit path.
- Use a jumper. If two traces must cross, you can jumper one across the other. Complete one of the traces, then draw the other trace up to where it crosses the interfering trace on both sides. Terminate each end of the blocked trace with a pad. When you build the board, solder a short length of insulated wire between the pads. The jumper wire "jumps" across the trace and makes the connection.
- Use a double-sided board. If a single-sided pc board requires many jumpers, consider using a double-sided board instead. Double-sided boards have many more routing options and should require no jumpers.

Laying traces on a double-sided board

On a double-sided board, one side typically contains mostly horizontal traces, while the other side contains mostly vertical traces. Figures 4-22 and 4-23 show examples of such a board. When a trace needs to change direction, a via routes the trace to

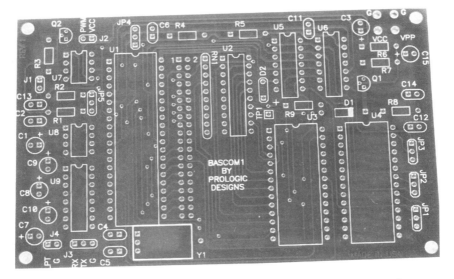

Fig. 4-22   *The top of a double-sided board, with most traces routed parallel to the long edges of the ICs.*

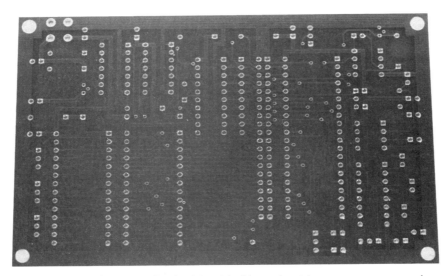

Fig. 4-23   *The bottom of a double-sided board, with most traces routed perpendicular to the long edges of the ICs.*

the opposite side of the board. The trace then bends 90 degrees and continues until the next direction change. Using this arrangement, all traces can be routed, although components might need to be pushed aside to make room for the traces.

Strictly following the technique of horizontal traces on one side and vertical on the other will not result in the simplest layout possible, but it can be a good starting point. When all connections have been made, you can remove unnecessary vias by shifting trace segments to the other layer when possible.

When you are drawing double-sided artwork, use different colors for different layers to avoid confusion. Many pc board design programs do this automatically.

For vias, be sure to provide a pad on both the top and bottom layers. Instead of a separate via, you can sometimes use a component lead hole to connect the top and bottom layers. If you use component pads as vias, you must be sure that you can solder to the pads on both sides (unless you are using plated-through holes). Resistors and axial-lead capacitors present no problems. But on many DIP IC sockets, the body of the socket covers the component-side pad and prevents soldering to it. Figure 4-24 shows a solution in the machine-pin collet socket, which is supported slightly above the board, allowing its pins to be soldered on both sides. Other solutions include eliminating the socket entirely and soldering to the component legs, or adding a short trace from the component pad to a separate via.

Fig. 4-24 *A machine-pin collet socket allows soldering on both the top and bottom sides of the board, so the socket pins can be used for interlayer connections.*

Multilayer artwork

In multi-layer artwork, the board layers might contain ground and power planes, where traces to ground and power supplies terminate. The ground and power planes are copper areas with non-conductive clearances around any vias that pass through the plane to another layer. A *blind via* is a via that runs from an outer layer (top or bottom) to an inner layer. A *buried via* connects inner layers only and does not appear in the top or bottom artwork.

Circuits with special requirements

By working carefully and following the guidelines in this chapter, you can achieve good results with many types of circuits. But pc board design is a complete field of study in itself, and the information in this book is not exhaustive.

If your circuits have critical timing, noise, spacing, or other requirements, you might want to learn more about pc board design techniques. Appendix C lists several books that provide more detail about designing pc board artwork.

If you don't want to tackle the pc board design for a complex circuit yourself, you can hire a professional designer to do it for you. Some pc board fabrication companies offer design services or will recommend an outside source.

Supplemental documentation

In addition to the main board artwork, there are other types of board documentation needed by some pc boards. Every board should have a parts placement diagram. Some boards also require one or more of the following: solder mask artwork, drill or fabrication drawing, photoplotter file, and drill file.

The parts placement diagram

Along with your board's artwork, you'll want to prepare a parts placement diagram, or component assembly drawing, which shows the locations and labels for all components on the board. You can use the diagram as a guide when you insert components into the board. It's also a valuable reference for component identification during board testing.

Commercially fabricated boards often include parts placement information printed on the board. Because silk screen printing was traditionally used for this process, the pattern to be printed is often called the silk screen artwork, whether or not silk-screening is actually used.

For prototypes and small-scale fabrication, the on-board parts placement diagram might be unnecessary. Limited information, such as a board label and polarity indicators, can be etched on the board as part of the artwork.

Whether or not the parts placement information is printed on the board itself, be sure to document your boards with a diagram that shows all components and their placement on the board. The diagram should label each component by number, corresponding to the numbering on the schematic and the parts list (R1, R2, U1, U2, and so on). The diagram should also indicate polarities of ICs, diodes, electrolytic capacitors, and other components with required orientations. For some transistors and other devices, an outline of the package in the diagram indicates the correct orientation.

Some CAD software generates a parts placement diagram automatically, using the information in the netlist and board layout. Otherwise, you can draw the diagram separately. Some software allows you to draw the parts placement diagram along with the pc board artwork using different screen colors and a different drawing layer for the parts placement information. This technique makes it easy to place the part outlines correctly in the drawing.

When possible, an effective technique for printing the parts placement diagram on paper is to print the board artwork (the pads and traces) in light gray, with the parts placement information superimposed on the artwork in black, as shown in Fig. 4-25. If your software doesn't allow this, you can print out the parts placement information by itself, or even draw it by hand on a separate sheet of paper. If you draw it by hand, use graph paper for best results.

Solder mask artwork

On commercially fabricated boards, all areas except those that will be soldered are covered by a protective coating, traditionally green in color. This coating is called the solder mask, and it might be applied as a dry film or screen printed in liquid form onto the board. The solder mask protects the artwork and helps to prevent solder bridges between traces and pads during soldering.

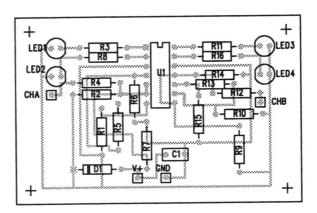

Fig. 4-25 *A parts placement diagram that shows the artwork in gray for easy reference.*

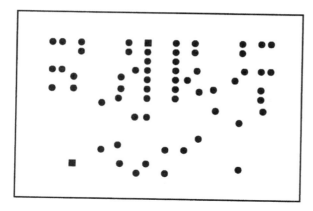

Fig. 4-26 *Solder mask artwork.*

The solder mask artwork is a to-scale drawing of the board indicating which areas should be coated by the mask and which should be left bare (Fig. 4-26). Prototypes and other hand-fabricated boards often omit the solder mask, making this artwork unnecessary for these boards.

Solder paste mask

On surface-mount boards, solder paste might be screen printed or stenciled onto component pads in preparation for component placement and soldering. This requires solder paste mask artwork, which identifies the areas where solder paste is to be applied on the board.

Blueprint or fabrication drawing

The blueprint shows a board's dimensions, hole locations, and other special instructions for board fabrication. The hole locations might be used in preparing a drill tape for automated drilling, although many software programs now generate drilling files directly. The blueprint is optional for board fabrication by hand. Figure 4-27 shows an example of a drill drawing.

Other documentation

In addition to the artwork, there are other types of documentation used in pc board fabrication, especially when the board is designed with the aid of a computer and the board will be fabricated commercially. Two files generated by many CAD programs are drill and photoplot files.

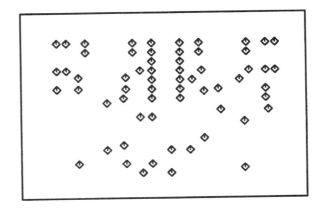

SIZE	QTY	SYM
37	62	◇

Fig. 4-27 *A drill drawing.*

A *photoplotter* is a machine that draws precise, accurate lines and shapes in light on a photosensitive medium. Figure 4-28 shows a photoplotter from the Gerber Scientific Instrument Company. A Gerber *photoplotter file* is a text file that describes the locations and sizes of all pads, traces, and other artwork elements for use with a Gerber or compatible photoplotter.

Fig. 4-28 *A photoplotter draws the artwork in light on a photosensitive medium. (Gerber Scientific, Inc.)*

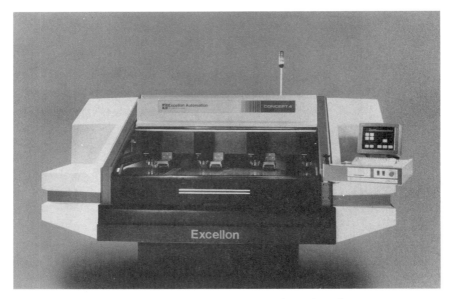

Fig. 4-29 *An automated drilling system for pc boards. The computer terminal on the right gives an idea of the size of the machine. (Excellon Automation)*

The photoplotter reads the information in the file and uses it to determine where to plot. The finished plot is developed photographically, and the result is a high-quality image of the artwork that is ready for use in exposing a sensitized pc board. The accuracy of photoplotted artwork is excellent, as good as ±0.0002 inch.

A *drill file* is a text file that describes the location, size, and depth of every hole to be drilled in the board. The drill file is read by a computer-controlled drilling system, which uses the information to drill the holes automatically. Figure 4-29 shows a drilling system from Excellon Automation Company. Positioning accuracy with these machines is better than ±0.0002 inch, which is far better than you can achieve with hand-drilling.

Artwork final check

The time to check your artwork for completeness, accuracy, and good design is before you transfer the image to the board. When you think your artwork is complete, answer the following questions:

- Does the artwork include all of the on-board components?
- Are all the components oriented correctly?

- Does the artwork include all of the connections in the circuit, including those to off-board components?
- Are all of the connections included and routed correctly?
- Are pad sizes, shapes, and spacing appropriate?
- Are traces and spacing wide enough for easy fabrication?
- Will all components fit in their assigned places on the board, with no overlap or interference?
- Are unused component leads terminated, if necessary?
- Are board and component mounting holes included, if necessary?

When you can answer "yes" to each of these questions, you're ready to transfer your artwork to the pc board.

5
Drawing the
pc board artwork

When you are ready to design your pc board artwork, you can choose from a variety of techniques, both manual and computer-aided. Each method has benefits and drawbacks. Which you choose depends on your budget, the equipment and materials you have on hand, the circuit complexity, and how many copies of the board you need.

This chapter describes techniques for creating pc board artwork, including how to use press-on and rub-on transfer patterns, and how to use pc board design software and general-purpose CAD software. The buyer's guide will help you choose software for pc board design. If you are using existing artwork, such as the patterns published in many hobby or electronics books and magazines, the design work is done for you, and you can move on to transferring your artwork to the pc board.

Transfer tapes and patterns

A long-popular, low-tech, and low-cost way of laying out artwork is to use press-on or rub-on transfer patterns and tapes that are designed for this purpose. Until computer-aided design came along, these methods were the best available. Today, complex circuits are more likely to be laid out with the aid of a computer, but transfer tapes and patterns are still a workable method, especially for simpler circuits.

There are two main types of transfer patterns: press-ons and rub-ons, with the names describing their method of application. Which to use is a matter of personal preference. Experiment with both to find which you prefer. If you wish, you can use a combination, using rub-on pads and press-on traces, for example.

Step-by-step: Using press-on tapes and patterns

Figure 5-1 shows examples of press-on patterns and tapes. The press-on patterns are opaque, adhesive-backed shapes that you lift from a backing sheet and press onto the artwork sheet. They are available in all common pad sizes and shapes. Multiple pads for an entire IC, transistor, or other component might be printed on a single common carrier sheet that allows you to place all of the pads in one step. Circuit traces are formed from an opaque, flexible, adhesive-backed tape that you unwind from rolls and cut to size. Available tapes range from 0.015 to 2 inches in width.

The press-on materials can be used with several methods of image transfer. You can even apply the patterns directly to a pc board, where they function as the etch resist. Chapter 7 describes this process.

The patterns are opaque and can be placed on a transparency for use in photographic methods of image conversion and transfer. This includes contact printing of the image onto a pc board, using reversing film to create a negative from a positive image, and photographing the artwork with a camera for later enlarging and exposing onto a pc board. You can also photocopy taped artwork for use in iron-on image transfer.

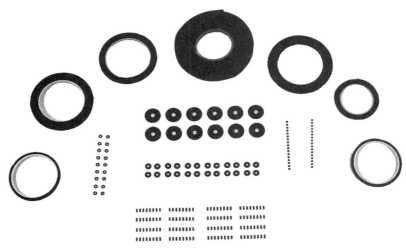

Fig. 5-1 *Press-on pads and tapes are available in many sizes, including ×1, ×2, and ×4 scales.*

The following paragraphs describe how to create artwork on a transparency using press-on patterns.

Materials

You will need the following materials (Fig. 5-2):

- A sketch of the artwork to be created, preferably on graph paper, drawn to the desired scale.
- An assortment of patterns and tapes. If you will be photographing the artwork, decide whether you will create the original artwork at ×1 (actual size), ×2, or ×4 scale. Be sure you have pads and tapes in the sizes you need. For example, if you are working at ×2 scale, a pad that will be 0.04 inch in diameter on the pc board requires a 0.08-inch pad in the original ×2 artwork.
- A sheet of transparent polyester drafting film. Polyester drafting film is dimensionally stable and will not stretch or shrink with changes in temperature or humidity. It's available at art, engineering, and office supply stores. Mylar is a popular polyester manufactured by duPont. An antistatic finish will repel dust. If you will be photographing the artwork, a matte finish on the film will reduce reflections.
- An artist's or craft knife.
- A burnisher. This can be any hand tool with a hard, smooth surface, such as the back of a plastic comb.
- Masking or other tape to hold the film on the artwork sketch.

Work area

Work at a drafting table or desk, and use a task lamp to illuminate your work surface.

Procedure

Follow these simple steps:

1. Cut a piece of polyester film slightly larger than your artwork—leave about a ½-inch border on each side. Tape the film over your artwork sketch, applying tape only to the edges of the film.
2. Begin the layout by defining the shape of the board. Press a line of tape along each outside edge of your artwork, or at least at the corners, to define the board's boundaries.

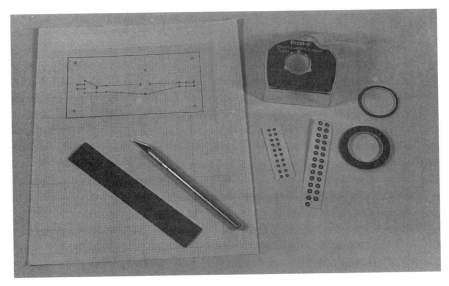

Fig. 5-2 *Materials required for creating an image with press-on patterns.*

To lay the tape, lift the tape from the end of the roll and press it onto the polyester film at the desired starting location. Unroll the tape, pressing it into place as you do so (Fig. 5-3). Don't stretch the tape as you place it. When you reach the end of your desired line, carefully cut the end of the tape with an artist knife (Fig. 5-4).

3. When the board's shape has been defined, you're ready to create the artwork. You might develop your own preferences for what order to follow in laying down the pads and tapes, but here is one method that works.

First, lay the pads. Begin with the IC pads, if any. For these, you can lay individual pads one by one, or you can use patterns that consist of a set of prespaced pads on a common carrier sheet. These allow you to lay perfectly spaced pads for a multileaded component in one step. Whichever pattern type you use, use the tip of an artist knife to carefully lift the pattern from its backing paper and position the pattern on the transparency (Fig. 5-5). Try to position the pattern accurately the first time. But if you do make a mistake or change your mind, you can lift and move the pattern.

4. After placing the IC pads, move on to the other component pads. If you are working at ×2 or ×4 scale, remember that all pads, traces, and spacings must be at the same scale.

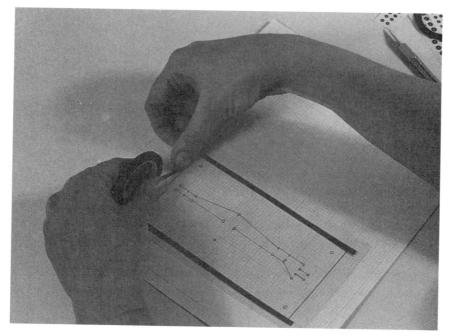

Fig. 5-3   *Laying tape to define the board edges.*

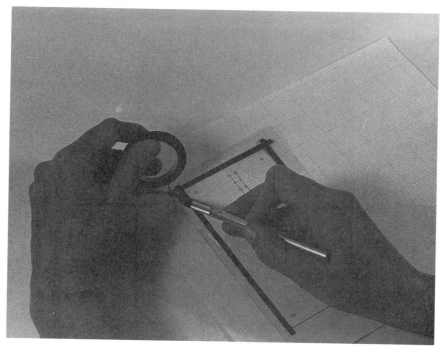

Fig. 5-4   *Cutting the tape.*

Fig. 5-5 *Laying pads.*

5. Lay the interconnecting traces. Press one end of the layout tape so that it overlaps the desired pad, but not its center hole. Press the tape along the transparency until it meets its interconnecting pad. Don't stretch the tape as you lay it. At the point where you want to cut the tape, hold the blade of the artist knife horizontal to the artwork sheet, and cut the tape so that it overlaps the pad it connects to, but not the center hole. Be careful not to cut into the pad as you do this (Fig. 5-6).

   You can make curved or angled corners with the tape. To make a curved corner, bend the tape in a smooth curve as you lay it. The tape is somewhat flexible and will form curves. To make a right-angle corner, you can lay two separate traces that intersect, or you can lay the corner with a single piece. To do this, lay the first segment up to the desired corner. Then cut the tape most of the way, but not all of the way through, bend it, and lay the second segment at the desired angle. The cut allows the tape to bend without stressing it too much.

6. If you are creating your artwork at ×2 or ×4 scale, add at least three accurately placed reduction targets for use in creating the actual-size artwork. Figure 5-7 shows an example. The targets are placed near the corners, but outside

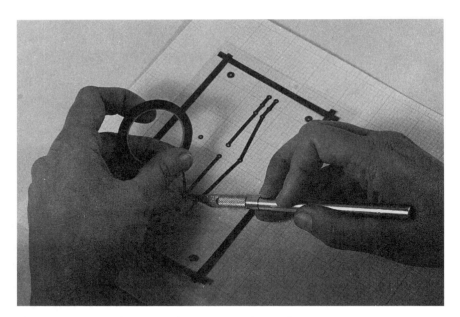

Fig. 5-6   *Cutting the tape where it joins the pad.*

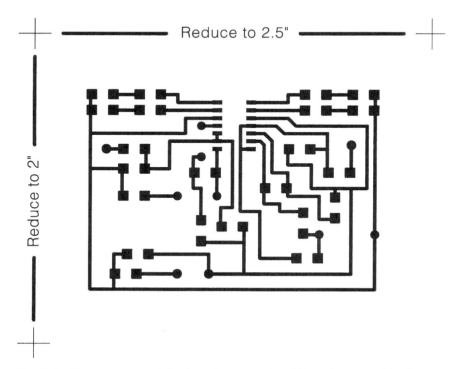

Fig. 5-7   *Targets are useful when the artwork will be photographically reduced.*

of the board artwork. The targets indicate the proper scale for the artwork. For example, if the artwork is drawn at ×2 scale and the targets are 6 inches apart, on the actual-scale artwork the targets would be 3 inches apart.

7. When you think you've completed the layout, double-check to be sure all connections and components are accounted for, and that all are correct.

8. Burnish the artwork to ensure that all pads and traces adhere to the transparency. Lay a sheet of paper over the artwork and rub over the pads and traces with a burnishing tool, the back of a plastic comb, or any hard, smooth surface (Fig. 5-8). This helps to ensure that the patterns adhere tightly to the film.

9. Label the artwork with at least the project name and date. This will save you time and trouble later, when you are searching for or trying to identify a particular piece of artwork. Store the artwork carefully to keep it from being damaged.

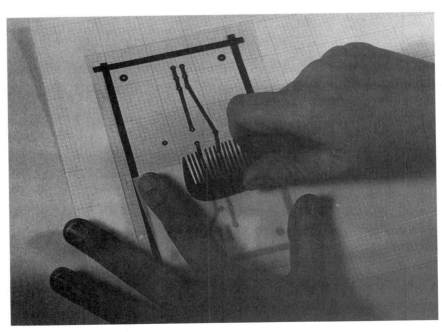

Fig. 5-8 *Burnishing the artwork onto the film.*

Step-by-step: Using rub-on patterns

Rub-on transfer patterns are similar to the press-on patterns described previously. The patterns, which include pads and traces, adhere to a semitransparent carrying sheet. Instead of lifting the patterns off a backing paper and pressing them onto a transparency, you rub the patterns directly from their backing paper onto the transparency. Figure 5-9 shows examples of these patterns. The process for creating artwork with rub-on transfers is similar to that used for press-on transfers.

Materials

Use the same materials and work area listed previously for press-on patterns, substituting rub-on patterns for the press-ons.

Procedure

Follow these simple steps:

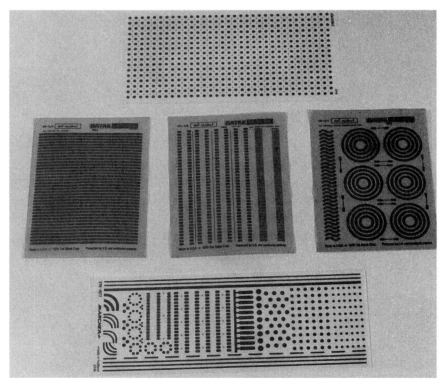

Fig. 5-9 *An assortment of rub-on transfer patterns.*

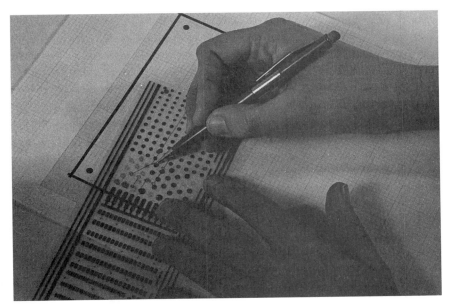

Fig. 5-10   *Transferring a pad onto the artwork.*

1. As with the press-on patterns, transfer the IC pads first, then the other pads. To transfer a pad to a transparency, place the pad in its desired position over the transparency, transfer side down. You can easily identify the transfer side because you can cut or scrape the patterns with a knife. The backing sheet is usually matte or semitransparent. Using a pencil or ballpoint pen, rub over the desired pad on the carrying sheet (Fig. 5-10). This frees the pad from the carrying sheet and causes it to adhere to the transparency.

2. Transfer the traces. The traces are straight lines that run the length of the carrying sheet. To transfer a trace, first cut the trace to size. Lay the carrying sheet, transfer side up, in the desired position on the artwork and use an artist knife to carefully cut a portion of the trace to the desired length. Cut the trace so that it overlaps the pads it connects to, but not any holes in the pads. Be careful to cut only the transfer, not the backing sheet. Flip the backing sheet over, position the trace in its desired position on the artwork, and rub over the trace to transfer it to the artwork (Fig. 5-11).

3. If needed, lay reduction targets.

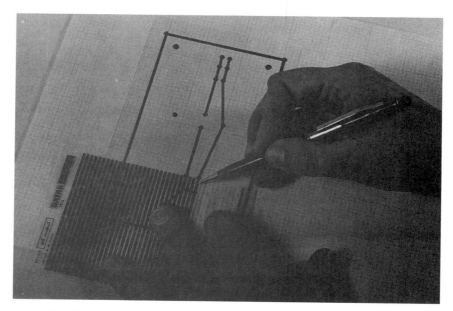

Fig. 5-11   *Transferring a trace onto the artwork.*

4. When all the patterns have been laid, check your work and burnish as described previously.
5. Label the artwork. Store carefully.

Creating double-sided artwork without a computer

Double-sided boards require at least two separate pieces of artwork: one for the top and one for the bottom of the board. With two pieces of artwork, you must be sure that all of the pads that are common to both layers have accurate registration, with all top and bottom pads lining up precisely, one on top of the other.

An alternate approach to making double-sided boards is to create three sheets of artwork. The top and bottom sheets each include the traces and pads that appear on one side only, while a third sheet, called the *padmaster,* includes the pads that appear on both layers.

To print, photocopy, or expose the top artwork, the top artwork and padmaster are laid one atop the other to form a single image. To do the same for the bottom artwork, the padmaster is flipped over, creating a mirror image that corresponds to the orientation of the bottom artwork.

The advantage of the three-layer method is that you only have to lay the common pads once, and because the same pad-master is used for both layers, the pads on the top and bottom artwork are guaranteed to be identical.

Yet another approach for creating artwork for double-sided boards uses photography and filters. The artwork is prepared on a single transparency, with different-colored tapes for the top, bottom, and common artwork. You then photograph the artwork twice, using special camera lenses and filters that selectively make the top and bottom colors invisible.

For example, you can lay the top traces in red tape and the bottom traces in blue, with black pads that appear on both sides. With colored filters, the camera photographs the red and black artwork with the blue traces invisible, then the blue and black artwork with the red traces invisible, to create two artwork images for transferring to the two sides of the pc board.

Computer-aided design for pc boards

As with schematic drawing, a personal computer and software can help you create your pc board artwork. In fact, computers are especially useful in this step because they can create precise, accurate artwork images without the tedious handwork of laying pads and traces. Some software will even place the components on the board and route the traces for you. Many programs work along with schematic drawing packages to ensure that the artwork matches the schematic.

As with schematic drawing software, software for pc board layout is available at all price levels and many levels of sophistication. The software must be able to store the artwork's image in a form that can be used to print, plot, or draw accurately scaled artwork. Software for designing pc board artwork includes products designed specifically for this purpose, as well as general-purpose CAD software that you can use for mechanical, architectural, or other drawing.

Any of a number of output devices can draw the final artwork. A dot matrix printer can print the artwork on paper, a laser printer can print on paper or a transparency, and a pen plotter can plot on paper, transparency, or even directly onto copper. The computer might generate a special file for use with a photoplotter, which draws the artwork in light on a photosensitive medium.

Creating double-sided artwork with a computer

Creating double-sided artwork with a computer is much easier than doing so by hand. Most software can create a drawing on multiple layers, with one layer holding the top artwork and another holding the bottom artwork. Multilayer boards use additional layers.

Typically the computer screen can display one layer at a time, or both layers at once, with different colors to represent artwork that appears on the top only, bottom only, or both layers. To add a via, you need only select the desired location on the board and the software places pads on both layers automatically.

When it's time to print or plot the artwork, the software can split up the layers and print each individually, mirror-imaging one of the layers if necessary.

Software for pc board design

As with schematic drawing software, many products are available for designing pc board artwork. Many of them work along with the schematic drawing programs described in chapter 2. Products range from basic, low-cost or even free software to higher-priced, more full-featured, professional versions.

Step-by-step: Designing
pc board artwork with PADS-PCB

PADS-PCB from PADS Software is an example of dedicated software for designing pc board artwork. PADS-PCB is the companion to PADS-Logic (described in chapter 2). As with PADS-Logic, both professional and free evaluation versions are available. The evaluation version is limited to circuits of about 30 ICs.

The following sections describe the PADS-PCB software in greater detail. The procedure described is just one example using one software product. Other products can perform the same functions, but vary in their specific capabilities and how they are implemented.

Figure 5-12 shows the main drawing screen of PADS-PCB. It's similar to the drawing screen for PADS-Logic, with the working area taking up most of the screen, the system information window and command menu along the left, and the prompt line on the bottom.

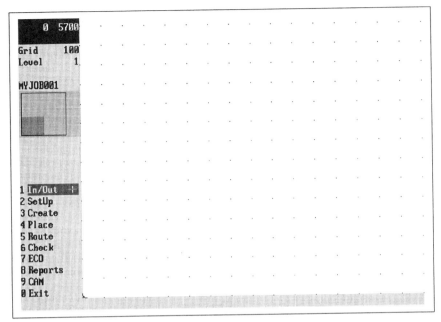

Fig. 5-12   *The main drawing screen of PADS-PCB.*

The command menu for PADS-PCB is similar to the command menu for PADS-Logic, with some differences because of the different functions that PADS-PCB performs. The functions of the items in the main command menu for PADS-PCB include

- *Job In/Out.* For loading artwork and netlists, and saving artwork.
- *SetUp.* For selecting display colors and preferences, and setting minimum clearances, routing levels, and other board parameters.
- *Create.* For creating a board outline, copper area, line drawing, or new part.
- *Place.* For moving and placing components.
- *Route.* For routing traces.
- *Check.* For checking for spacing violations, crossed traces, other problems.
- *ECO* (engineering change order). For revising artwork and schematics.
- *Reports.* For generating netlists, parts lists, and statistics about layout quality.
- *CAM* (computer-aided manufacturing). For printing, plotting, and creating files for photoplotters and numerically controlled (N/C) drills.

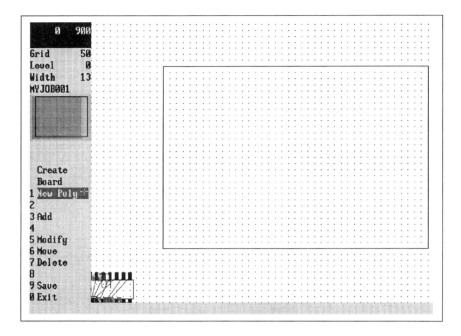

Fig. 5-13 *Defining the pc board. In the corner are components loaded from a netlist.*

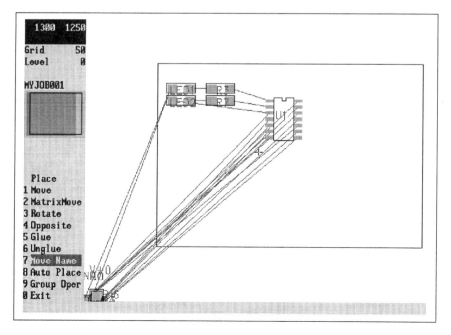

Fig. 5-14 *Placing components on the board.*

As with PADS-Logic, you can select items from the command menu by using the designated function keys or by moving the cursor into the command menu area and clicking the mouse.

From the numeric keypad, you can zoom in or out to view any area, from the entire drawing to any section you select with the cursor.

The following steps illustrate how to create pc board artwork using PADS-PCB. They describe a procedure that works for me. You might develop different preferences about the order of parts placement, routing strategies, and other details of pc board design. The procedure assumes that the software is installed and configured to match your hardware and preferences, and that you have created a schematic and netlist for your circuit with PADS-Logic.

1. Run PADS-PCB. This calls up the main screen and displays the working area and main menu.
2. Load your netlist into the program. Select *In/Out,* then *ASCII In,* and enter the file name of the netlist you created with PADS-Logic or another schematic capture program. PADS-PCB reads in your netlist and places the component outlines in a pile in the lower left corner of the drawing area.
3. Define the pc board. From the main menu, select *Create,* then *Board,* then *New Poly.* Use the cursor to draw a rectangle that matches the size of your pc board (Fig. 5-13). The cursor-position numbers at the top of the screen show distance in mils, or thousandths of an inch. To draw a board 1.5 by 2 inches, move the cursor 1500 mils, select *Add Corner,* move the cursor 2000 mils, select *Add Corner,* repeat these steps, and select *Complete* to draw the final edge.
4. Select a routing grid. All traces and components will be placed on a grid, whose size you can define. To set the grid to 0.05 inch (50 mils), type *G50* on the command line.
5. Select a trace width. To set the width of the traces you will draw to 0.020 inch, type *W20.* You can change the width whenever you want.
6. Place major components, including any components with fixed locations, on the board. Figure 5-14 shows beginning parts placement for a pc board using surface-mount components.

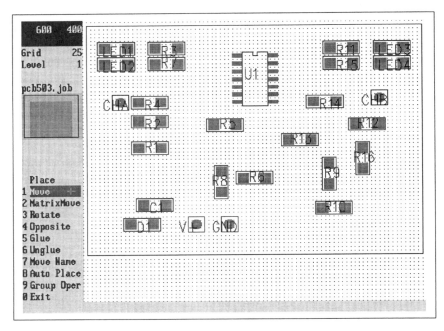

Fig. 5-15 *Component placement, with routing invisible.*

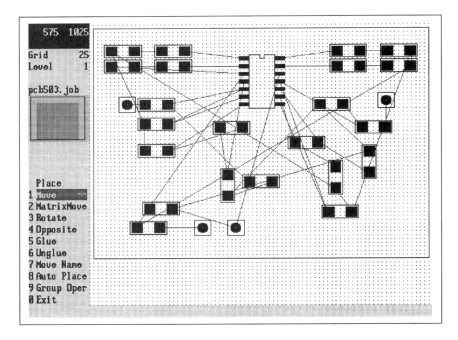

Fig. 5-16 *Rat's nest routing, with straight-line, pad-to-pad connections, ignoring routing constraints.*

When the components are first loaded into the drawing, they are "glued" onto the screen. Before you move a component, you have to unglue it. To unglue, select *Place,* then *Unglue.* You can unglue all components by typing *, or unglue all resistors by typing *R**, or unglue a single component by typing its designation: *R1,* for example. For this example, we'll unglue all of the components at once.

To place a component on the board, select *Place,* then *Move* from the main menu. You can place the cursor over the IC's outline and select *Select,* or you can type a component designation. When a component has been selected for moving, the outline of the part moves as you move the cursor. Attached to the part are "rat's nest" lines that indicate connections to other components in the circuit. These will be used later, when routing traces. Move the component to its desired position on the board (you can change the position later if necessary), and select *Complete* to place it on the board.

7. Place the other components. Continue to select components and place them in their estimated locations on the board. The rat's nest lines can help you decide which components should be placed near each other on the board. The *SetUp* menu enables you to choose whether or not to display part outlines, component designations, connections, and other information. Figure 5-15 shows all components on the board, with connections invisible. Figure 5-16 shows component placement with the unrouted, rat's nest connections visible.

8. Begin routing the connections. From the main menu, select *Route.* Move the mouse over one of the connection lines and select *Route Connection.* As you move the mouse, the connection line moves with it. On one side of the cursor, the connection changes color to indicate that this is the segment you are routing. You can move it up, down, left, right, or at a 45-degree angle from the pad it connects to. When the segment is in the desired position, select *Add Corner* to continue routing in a different direction, or *Complete* to finish routing the trace on the board. Figure 5-17 shows a partly routed board.

9. Continue routing the traces, and modify them as necessary to complete or improve the routing. To change a trace, go to the *Route* menu and select *Modify* to add, delete, or move a corner in a trace, *Reroute* to move a segment in a

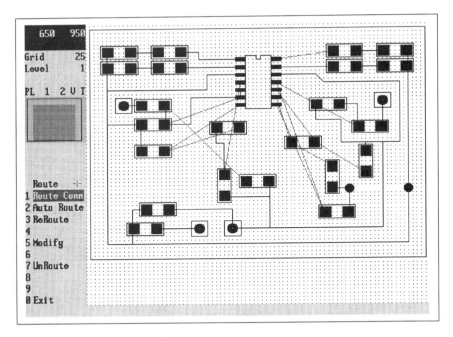

Fig. 5-17 *A partly routed board.*

trace, or *Unroute* to convert the trace back to an unrouted
rat's nest connection so you can start over.

You can also move components using the same commands you used to place the components on the board. If
you move an already routed component, its traces will
follow along with the component pads, stretching or
shrinking as necessary. After moving an already routed
component, you might need to move the traces for a
neater or improved routing. Figure 5-18 shows a group
move in progress, with the selected group of components
surrounded by a box.

10. Check the artwork for completeness and accuracy. On request the software will check to see if any of the pads or
traces violate the spacing requirements you specify (Fig.
5-19). Figure 5-20 shows a completely routed board.

11. Print or plot the artwork. To print the artwork, select
CAM, then *Direct,* then *Matrix Printer* or *Laser Printer,* as
appropriate. Figure 5-21 shows the items you can select
for printing.

These are the basic functions available in PADS-PCB. Many
more are available, including autoplacing of components on the

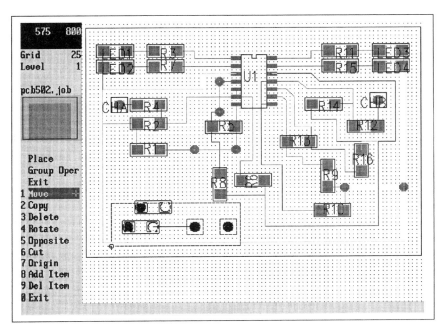

Fig. 5-18 *The components in the box are being moved as a group.*

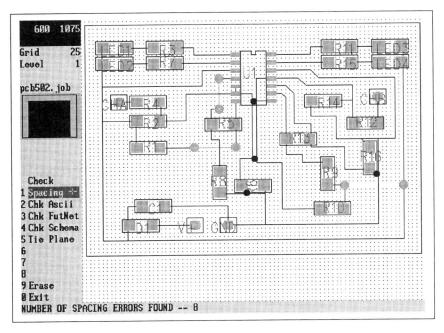

Fig. 5-19 *On request, spacing errors are identified.*

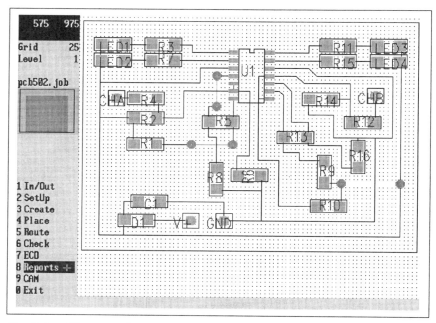

Fig. 5-20   *The complete, routed board.*

16004 16107		Select Items		Laser Printer			
				General Plot			
ITEMS			ITEMS	L1			
Board	4		Pads	1			
Connections			Vias	1			
Parts - Top	4		Tracks	1			
Parts - Botm			Copper Areas				
Ref-Designators	4		Line-items				
Part Types			Text	4			
			Part Outlines	4			
Select Grey Level	4	3 2 1					
PROCEED		GO BACK		EXIT			

Grid 25
Level 1

pcb502.job

CAM
Direct
1 Select +
2 Proceed
3 Go-Back
4
5
6
7
8
9
0 Exit

Fig. 5-21   *Printing options.*

board and autorouting of connections. If you don't want to design your artwork from a netlist, you can design it "on the fly," adding each component and connection as you design the board.

Buyer's guide to pc board design software

If you are in the market for pc board design software, you should decide on what features you need and choose the best product you can find that fits your budget. The best way to become familiar with any pc board design software is to use it and explore its possibilities. As with schematic drawing software, many companies offer free demo versions of their software so you can get a feel for how the programs work before you buy. Appendix A lists a variety of software sources.

The features of pc board design software are discussed in the following paragraphs to help you decide what to buy. Some features, like the basic ability to place pads and traces precisely, are essential, though programs vary in their ease of use. Other features, like autorouting, are optional. In pc board design software, even more than in schematic capture software, the range of features and levels of performance are enormous. Prices range from free to thousands of dollars.

Overall requirements

Chapter 2 included questions to ask about system requirements, the user interface, and documentation and support for schematic drawing programs. These same questions also apply to pc board layout software.

Input options

Input options have to do with how the software receives information about the circuit.

- Can the software read netlists from my schematic capture program? If you use a schematic capture program, be sure that your pc board design software can read its netlists. This saves you from having to reenter circuit components and connections manually.
- Can I enter pads and connections manually without the aid of a netlist? Occasionally you might want to design artwork on the fly, without first creating a schematic and net-

list on the computer. If you want this capability, be sure your software offers it.

Placing pads

Programs vary widely in their ease of placing component pads on the board.

- Is the grid size variable? High-density artwork with closely spaced pads and traces requires a finer grid than lower-density boards. Grids as small as 0.001 inch allow you to place components precisely where you wish. Some programs also allow you place pads off-grid, for maximum flexibility.
- Are both metric and imperial (decimal inch) grids supported? Currently most component pads are spaced in decimal fractions of an inch (0.10 inch, 0.05 inch, etc.), but components with metric spacings also exist.
- Will the software place components automatically on the board? With autoplacing, the software analyzes the information in the netlist and places components on the board to give the shortest traces overall. This gives a starting point for routing the board, though you might want to move components later.
- Can I easily rotate components on the board? Many programs make it easy to rotate components 90, 180, or 270 degrees, allowing you to place components on the board at any orientation.
- Can I swap gates, pins, or other elements to try out different routings? Sometimes swapping gates or pins will solve a routing problem. Some software will handle this automatically on request, swapping all related connections in one step.
- How many components can I place on a board? Be sure your software can handle the most complex boards you need to design.
- Can I fill an area on the board with copper (for shielding or a ground plane, for example)?

Routing traces

In addition to placing pads, the other major task of pc board design is routing the traces. Programs vary widely in how much help they offer in this area.

- Can I view a rat's nest of unrouted connections? A rat's nest shows the connections in a circuit drawn as straight lines from pad to pad, without regard for routing rules. The rat's nest helps to show which components connect to which, and can serve as a guide for arranging components on the board for easiest routing.
- Can I use rubber-banding to see how the traces route as I place them? Rubber-banding, or viewing connections as you route them, is useful in pc board design as well as schematic-drawing.
- Is autorouting included? With autorouting, the software routes the traces automatically. Some software allows you to autoroute selectively, routing individual traces that you specify, or routing only the power and ground connections, for example.
- How capable is the autorouting? Autorouters vary greatly in what they can do and in how fast they do it. They might use any of a variety of algorithms, or procedures, for finding routes between pads. Not all autorouters can route 100 percent of every board. Some are better than others at finding the shortest routes or the routes with the fewest vias. Some allow more flexibility than others in specifying layout rules to follow, such as special spacing or placement requirements for sensitive components.
- How fast is the autorouter? There is wide variation here as well.
- Are pads automatically shaved or traces necked (narrowed) to allow adequate spacing when a trace runs between pads?
- Can I draw curves or corners at any angle? Simple boards might be routed with right-angle corners only, but some software will draw angled or rounded corners as well.

Component libraries

In many ways, the component libraries for pc board design are similar to those used for schematic capture software. In some products, a single library holds the symbols for both the schematic and the pc board. In other products, the two libraries are separate. Chapter 2 includes several questions to ask about component libraries for schematic capture software. The questions about library size, contents, ease of use, and new-component design also apply to pc board design software. In addition, the following questions relate specifically to libraries for pc board design:

- What package types are supported (DIP, SOIC, QFP, etc.)?
- Is it easy to change package types in a drawing? For example, some components are available in both DIP and SOIC packages. Some software allows you to select a different package type and the pads change automatically to match.
- Does the library include popular connector symbols (D-connector, edge connector, etc.)?
- What pad sizes and shapes are available (round, oval, square, rectangular, connector finger, and custom)?
- Can I draw through-hole pads with or without center holes? Center holes are helpful as drill guides for hand-drilling, but unnecessary if you are sending your boards out for fabrication.

Board limits

Your pc board design software should be able to handle the largest board you will create.

- What board sizes are available? Some products limit you to a few sizes, while others allow you to define the size and shape.
- How many layers are supported? This is important if you are designing boards that are double-sided or multilayer.
- Can I place both through-hole and surface-mount components in the artwork? In surface-mount circuits, the solder side and component side are the same. Software with full surface-mount support allows you to place a component on either the top or bottom, and the pads will be oriented correctly. In addition, it allows you to place two surface-mount components in the same location, one on the top and the other on the bottom, because there are no through-holes to interfere.
- Can I place text in the artwork? You can draw pc board artwork without using text, but you might want to include a board label, polarity symbols, or other text, especially if a component placement diagram is not printed on the board.

Editing tools

Editing is important in pc board design because the initial parts placement and routing will almost certainly not be the final one.

The easier it is to try out different ideas, the easier the artwork is to design.

- Can I move component pads after placing them? When you move a component's pads, the traces that connect to them should follow along, stretching or shrinking as needed.
- Can I edit a trace after placing it in the artwork? Some software makes it easy to select a trace, "lift" it from its current location, and reroute it without having to erase and redraw in two separate steps.
- Can I move blocks of components? You might want to move a group of components to allow room for other components or traces. When you move a block, the traces that connect to the rest of the circuit should stretch or shrink as needed.
- Can I make global changes to drawing features, such as trace width, pad size, and so on? What if you design a board with 0.02-inch traces, and later you want to widen them to 0.03 inch? With some software, this is easy to do and can be done in a single operation.
- Can I make changes in the circuit design and easily change the schematic to match? What happens if you are in the middle of designing your artwork when you realize that you want to add one component or connection, or trade a component for a different one? It's useful to have the ability to do so, and to automatically update your schematic and netlist with the new information. In a similar way, it's useful to be able to add schematic or netlist changes to a previously routed board so the artwork can be altered to match. These changes are known as *forward annotations* (to the pc board) and *back annotations* (to the schematic).

Automatic design checking

Design checking helps ensure that your artwork is complete, accurate, and functional.

- Does the software check for adequate spacing between pads and traces? Some software allows you to define the minimum acceptable spacing between pads and traces and warns you if your artwork violates your specifications.
- Does the software check the artwork for accuracy and completeness against the netlist? Some software warns you if the artwork doesn't contain all of the components and traces in the netlist or if any are misrouted.

Outputs

Your artwork is useless if you can't get it out of the computer and onto a board.

- What printers and plotters are supported? Be sure your software supports the printers and plotters you use.
- Can I generate output files for a photoplotter or N/C drill? If your board will be commercially fabricated, the fabricator might request files for controlling a photoplotter or automated drill. Some software includes a special photoplot viewer that shows how the artwork will look when photoplotted so that you can check for and correct mistakes or problems before you send the artwork out for photoplotting.
- Are any other file formats supported (AutoCad DXF, Postscript)? If you want to import or export your files from or to other software, be sure your pc board design software has a file format available that both programs can use.
- Does the software create parts placement diagrams, solder masks, solder paste masks, and drill drawings? If you require these, be sure your software supports them.
- Can the artwork be printed or plotted at different scales? Depending on your method of fabrication, you might want to print or plot your artwork at actual, ×2, or ×4 scale.
- Can I print or plot the artwork as a mirror image or negative image? Some image transfer methods require mirror-image or negative-image artwork. Mirror-image printing ability is common, but negative-image printing is not.

General-purpose CAD for pc board layout

As with schematic design, you can use general-purpose CAD software to design pc boards. And as with schematic drawing software, you want to use a program that stores the drawing as vector graphics rather than as bitmapped graphics. In fact, this is more important for pc board artwork because it should be as precise and accurate as possible for best results.

To use general-purpose CAD software to draw pc board artwork, you need software that can do the following:

- Draw and place lines and objects precisely and accurately to within 0.01 inch.

- Draw solid lines of desired widths (0.02 inch, 0.03 inch, etc.).
- Draw solid, or filled, objects (component pads).
- Create components, such as a pad or group of pads, that you can call up and place in a drawing.
- Print or plot the artwork to scale.

Other capabilities that aren't strictly required include

- The ability to create mirror images of the artwork, and
- The ability to create a multilayer drawing, where lines and shapes are drawn as if they were on transparent sheets laid one on top of the other.

Because general-purpose CAD software has many applications, it generally won't have features that are unique to pc board design. For example, you generally won't find the ability to read a netlist from a schematic drawing program.

As a rule, general-purpose CAD software does not have extensive component libraries, specialized output files for photoplotters or drills, or design checking. Many popular general-purpose CAD programs, such as AutoCAD, have add-on software that adds some of these capabilities.

Step-by-step: Designing pc board artwork with Generic CADD

Generic CADD, which was described in chapter 2 as an example of a moderately priced, general-purpose CAD software for schematic drawing, can also be used to prepare pc board artwork. Figure 5-22 shows an example.

Generic CADD comes with a few basic pc board symbols, and a more extensive library is available (at extra cost). Or you can design your own symbols. Because many boards can be designed using only a few basic symbols, creating your own symbols is not that difficult for simple designs.

The following steps illustrate how to draw a pc board layout using Generic CADD. Other CAD software perform the same basic functions, but vary in their specific capabilities and other details. The following procedure assumes that Generic CADD is installed and configured for your computer and that you are familiar with basic computer operation. It also assumes that a component library of pc board symbols has been installed and added to the menus.

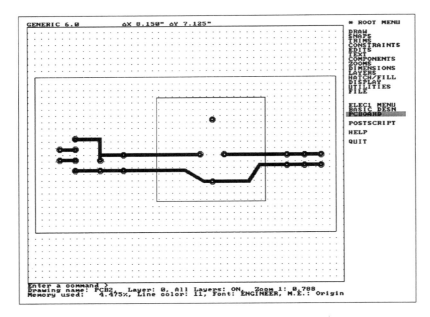

Fig. 5-22 *The pc board artwork drawn with Generic CADD.*

1. With the schematic for the board at hand, run Generic CADD. The main drawing screen appears.

2. Set the grid size to 0.05 inch or your preferred size. Set *Snap To Grid* on. This ensures that all pads, traces, and components are placed on grid points so you can easily and quickly draw interconnections that meet precisely.

3. Set the drawing limits slightly larger than your board size. Zoom in to view an area of about 4 by 5 inches.

4. From the pc board components menu, select components and use the cursor to place them in the drawing. Zoom and pan as required to view the area of the board you are working on (Fig. 5-23).

5. From the *Draw* menu, set the double-line options for your desired trace width. For example, for 0.2-inch traces, set the right and left offsets to 0.1 inch each. You can change the trace width whenever you want. Set *Solid* to *On* for solid, filled traces.

6. Using the double-line command, draw the traces (Fig. 5-24).

7. Save the drawing to disk.

8. Print or plot the drawing. You can print at actual scale or any other scale that will fit on your paper.

9. If you wish, you can also create a parts placement diagram and other artwork on additional drawing layers.

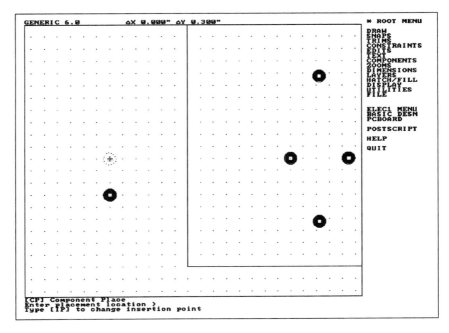

Fig. 5-23 *Zooming in to focus on one area of the board.*

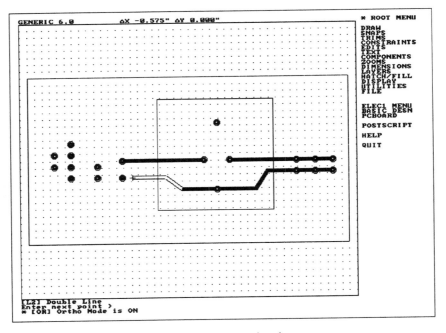

Fig. 5-24 *Using the double-line command to lay a trace.*

6
Preparing for image transfer

Before you can transfer your artwork to a pc board, you need to choose a board of an appropriate base material and size. You also need artwork that is in the correct format for use with your method of image transfer. Often you will need to convert your original artwork to a different form, whether it's converting from a positive to a negative image, copying an image on paper to a transparency, mirror-imaging the artwork, or reducing an enlarged image to actual scale.

This chapter includes information about how to choose the right board for your project, how to cut boards to size, and how to clean bare boards to ensure a good-quality image transfer. Following these are descriptions of a variety of methods for converting artwork from one format to another, and when to use each.

Types of pc boards

A pc board that doesn't yet have a circuit etched on it is called a *pc blank.* A single-sided pc blank consists of a base material, or substrate, with a thin layer of copper foil laminated or glued to it. A double-sided pc blank is similar, but with copper foil on both sides.

Multilayer boards have one or more internal, conducting layers. These boards are made by fabricating the individual layers and laminating them together with a bonding material to form a single board.

Fabricating multilayer boards generally isn't practical without special equipment to accurately and precisely register and laminate the layers. If you need a multilayer board, you can design the artwork yourself, but you will probably want to send the artwork out for board fabrication.

Base materials

Two broad categories of base materials for pc boards are paper-based and fiberglass-based boards. In each, thin layers of base material are bound together with a resin to form the base, and a sheet of copper foil is bonded to the base. Often you can tell the board type by its color, with paper-based boards having a brownish tint, and fiberglass boards having a greenish tint. Each type of substrate has advantages and disadvantages.

Paper-based boards are cheap and easily punched, but can expand and contract slightly with temperature changes. These boards can be difficult to drill without chipping or flaking. Paper-based boards are adequate for simple circuits, but are not recommended for surface-mount components or plated-through holes.

Boards with a fiberglass base are less likely to break, more stable thermally, easier to drill, and have higher insulating resistance. They are recommended for general-purpose as well as radio frequency circuits. These boards are more expensive than paper-based types.

There are many classifications of pc board base materials including

- XXXP. Paper base with a phenolic resin binder. The *P* indicates it can be punched by machine.
- FR-2. Paper base with a phenolic resin binder. Low cost, easily punched, and flame retardant.
- FR-3. Paper base with epoxy resin binder. Medium cost, good performance, and flame retardant.
- FR-4. Woven glass cloth with epoxy resin binder. Excellent electrical, physical, and thermal qualities. Drillable and flame retardant. Good general-purpose type.
- CEM-1. Paper and woven glass base with epoxy resin binder. Easily punched, with good electrical and physical properties.
- CEM-3. Woven and nonwoven glass base with epoxy resin binder. Easily drilled or punched. Good for plated-through holes.
- G-10. Woven glass substrate with epoxy resin binder. Similar to FR-4, but not flame retardant.

Board thickness

A typical pc blank is ⅟₁₆ inch thick, which is stiff enough to support components without flexing, yet thin enough to allow component legs to protrude through for soldering. Thicker boards are available to support large transformers or other heavy components.

There is an alternative to buying a double-sided pc blank. You can transfer your artwork to two thin (⅟₃₂-inch), single-sided blanks and align and laminate them together after etching. This technique is described in chapter 9.

Copper thickness

The copper layer on pc blanks is available in several thicknesses. Many boards use 1-ounce copper, which means that the sheet of the copper foil weighs 1 ounce per square foot. This works out to a thickness of 0.0014 inch. One-ounce copper gives good results with traces as small as 0.01 inch, or 10 mils.

For boards with very narrow traces, using a thinner copper layer might give better results because it will minimize undercutting during etching. As the etchant works its way vertically into the exposed copper, it also penetrates laterally, and etches away the edges of the copper underneath the etch resist, as shown in Fig. 6-1. The thicker the copper layer, the longer the etching time, and the more undercutting occurs.

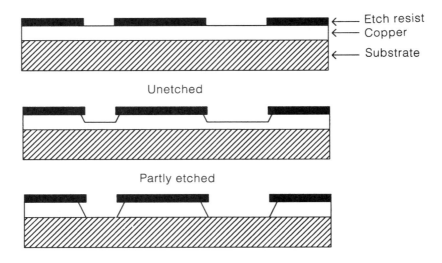

Fig. 6-1 *Undercutting occurs when the etchant begins to penetrate beneath the etch resist.*

Boards with thinner copper coatings are more expensive, and must use a high-quality base material like FR-4. With wide traces and pads, undercutting is less likely to result in a too-narrow or etched-through trace, and thicker copper can be used without problems.

Presensitized boards

If you will be using photographic methods of image transfer, you have to decide whether to buy pc blanks that are presensitized with photoresist, and if so, what type of photoresist to use. Chapter 7 describes the options in this area.

Cutting boards to size

Printed circuit board blanks are available in sizes as large as 24 by 36 inches or you can buy precut boards in a variety of smaller sizes. The precut boards are convenient, especially if you're making boards in small quantities, though they cost more per square inch than the larger boards. If you are using presensitized boards, buy precut ones, because cutting these boards without damaging their delicate coatings is difficult, if not impossible.

If you are making several boards at once, you can transfer the artwork for several boards (identical or different ones) to a single large pc blank, and cut it into individual boards after etching. With this technique, called *panelizing,* you can save time by transferring the artwork for several boards at once and etching them on a single board.

Panelizing is less practical if you are using taped artwork, where you can't easily produce multiple copies of the artwork. Also, the larger the board, the harder it is to achieve good-quality image transfer and etching of the entire board. In some cases, you might find it easier to cut the blank boards and do the image transfer and etching individually.

Cutting pc blanks to size requires a blade that will cut through the abrasive base material without dulling. Kepro offers pc board shears that cut boards cleanly in a single motion, much like a paper cutter. You can also use a hacksaw or circular saw to cut a board. Blades with carbide or diamond-steel teeth give the best performance and last longer, but a general-purpose saw blade will also do the job.

When using a hacksaw, clamp the board in a vise to hold it firmly during cutting, as shown in Fig. 6-2. To protect the board and keep it rigid as you cut, sandwich it between two blocks of

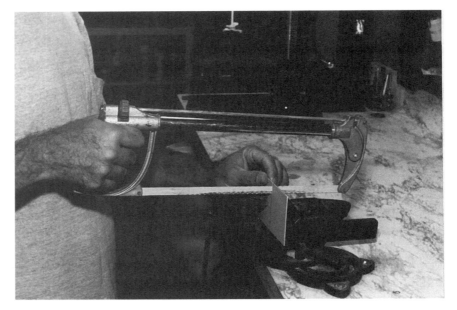

Fig. 6-2   *Cutting a pc blank to size.*

wood almost up to the cutting line. After cutting, smooth the edges of the board with a flat file (Fig. 6-3).

If you need to trim the edges of a board, it's better to do so before you transfer the artwork and etch. Cutting before etching conserves etchant, because you don't have to etch away the copper on board that will be removed. Also, if you damage the board during cutting, it's better to do so before you've put a lot of time into it.

Cleaning a pc blank

Transfer artwork only to a perfectly clean pc blank. Unless you are positive that the copper is free from contamination, you should clean the board before applying transfer patterns, iron-on artwork, or photoresist. Oil, dirt, grime, fingerprints, oxidation, and other contamination prevent the desired materials from adhering to the copper surface.

To clean a board, scrub it with a synthetic scouring pad and mild detergent or a bleach-free cleansing powder, and rinse it thoroughly with water, as shown in Fig. 6-4. Be sure to remove all detergent and other residues. When the board is clean, water should sheet evenly across the board's surface. If beads of water form on the board's surface, the board is not clean. You can also clean boards with isopropyl alcohol.

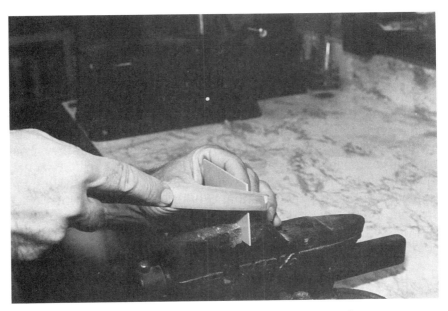

Fig. 6-3   *Using a flat file to smooth the edges of a pc board.*

Fig. 6-4   *Scrub a pc blank thoroughly before you transfer the artwork to it.*

After cleaning a board, don't touch its copper surface with your bare hands. Handle the board by its edges only (or by the bottom for single-sided boards). After cleaning, you can dry the board with a lint-free cloth or compressed air. Use the board soon after you clean it because delays will allow the copper to oxidize.

Converting artwork

Sometimes your original artwork isn't in the form you require for transferring to the pc blank. You might have to convert from one medium (paper, transparency, or computer disk) to another, from a positive image to a negative image, from ×2 or ×4 scale to actual scale, or from the original artwork to its mirror image. Here are some examples of situations that call for image conversions:

- A printout on paper from a dot matrix printer can't be used directly for iron-on image transfer because this method requires a dry-toner image created by a laser printer or photocopier.
- If you are using a pc board coated with a negative-acting photoresist, you need to create a negative image of your artwork.
- Artwork for a through-hole board drawn with a top-side orientation might have to be mirror-imaged for transfer to the bottom of the board.
- Artwork drawn at ×2 scale must be reduced to actual scale to transfer it to the board.

The following are the artwork formats required for each method of image transfer described in chapter 7:

- Transfer tapes and patterns—any drawing or sketch, positive image, ×1 scale, actual orientation (not a mirror image).
- Iron-on transfer—special iron-on transfer film or paper, positive image, ×1 scale, mirror image.
- Direct-plot onto copper—computer file, positive image, ready to print at ×1 scale, actual orientation.
- Contact printing—transparent film, positive or negative image (depending on type of sensitized board), ×1 scale, emulsion side of film should hold mirror image.
- Screen printing—transparent film, positive, ×1 scale, actual orientation.

The following sections describe ways of converting artwork from one format to another using a computer, photocopier, reversing films, and camera photography.

Computer methods of image conversion

If your artwork is stored on computer, your software probably allows some flexibility in how to print or plot a hard copy. Most CAD software includes functions to mirror-image a drawing. Most also allow printing at different scales, so you can print ×2 or ×4 artwork for later reduction with photography. If necessary, you might be able to divide a large artwork image into sections and print them individually, then tape them together for a single, large image. Figure 6-5 shows original artwork printed at ×3 scale, and the positive and negative actual-size transparencies created from it using photography. Software that converts a positive image to a negative is less common.

A dot matrix, laser, or other printer will print your artwork on paper. A laser printer can also print directly onto a transparency. Chapter 7 includes details on how to laser print onto iron-on transfer film. A pen plotter is another option for plotting on paper, transparencies, or even bare copper. For photography or photocopying, you can mirror-image the artwork printed on a transparency just by flipping it over.

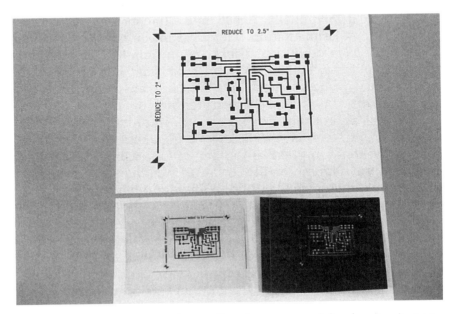

Fig. 6-5 *You can improve the quality of your artwork by drawing it at an enlarged scale, then reducing it photographically to actual size.*

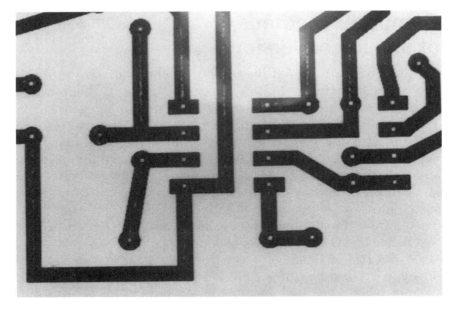

Fig. 6-6   *This laser printed artwork has voids, which appear as thin vertical holes in the traces. Voids can be prevented by increasing the printer's contrast, using a graphic arts toner cartridge, or switching to a different brand of paper or iron-on film.*

Printing a good-quality image onto a transparency sometimes takes some experimentation. The ideal image has solid, opaque traces and pads on a clear background. Some toner cartridges and transparencies produce solid areas that have tiny voids, which usually show up as thin, light or transparent lines that match the direction the paper feeds through the printer (Fig. 6-6).

Several approaches can be used to improve image quality. Most laser printers have a darkness adjustment that adjusts the amount of toner applied to the paper. Increasing the darkness might improve image quality. To avoid wasting toner, return the adjustment to its previous position for your other printouts.

Transparencies from different manufacturers vary in how well they accept laser printer toners. Experiment with different products until you find one that works well with your printer. Whatever you do, use only transparencies that are rated for use in laser printers. Transparencies intended for use with overhead projectors are not as heat resistant, and using them in a laser printer might damage or ruin the printer.

Finally, for best results, install a toner cartridge with an ultra-black, high-density, graphic arts toner. This toner is designed to print areas of solid black. In contrast, ordinary toner is designed primarily for printing the thin lines that make up text. Ultra-black toner cartridges are available from Black Lightning and other suppliers.

For photographic methods, instead of using a transparency, you can print onto *vellum,* which is a semitransparent drafting paper. Exposing a board through vellum requires at least twice the exposure time of exposing through transparent film, and the resulting image quality might be slightly less because of the longer exposure.

For iron-on transfer methods, there are special transfer films and papers available that can be used in laser printers, or you can use any transparency rated for use in your printer. Chapter 7 has more on this.

Photocopying

A photocopier is an easy way to copy artwork that is printed on paper, including book or magazine pages, onto a transparency or iron-on transfer film. Full-size copiers give better results than inexpensive desktop models. Be sure to use a transparency rated for use in photocopiers. Adjust the copier's contrast to give an image with opaque black and clear transparent areas.

A photocopier can also create a mirror image from a paper printout. Copy the image to be mirrored onto a transparency, flip the transparency over, and recopy it onto another transparency or paper.

Some photocopiers can reduce a ×2 or ×4 image to actual size. Check the resulting image carefully to be sure it's scaled accurately in both dimensions.

When using photocopiers, try to get the darkest blacks you can against a transparent or white background. The quality will deteriorate if you make several generations of a copy—making a copy of a copy of a copy, for example—so try to keep the number of generations to a minimum.

Step-by-step: Using reversing film

With reversing film, you can make a negative transparency from a positive transparency, or a positive from a negative. Exposed and developed areas on reversing film are an orange color that allows visible light to pass, but blocks ultraviolet frequencies. Unexposed areas on the film are clear and transparent to both ultraviolet and visible light.

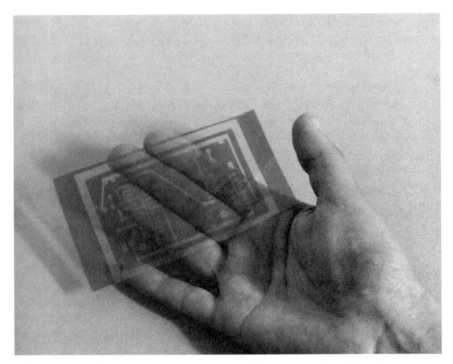

Fig. 6-7 *On this negative image exposed onto reversing film, the orange background allows visible light to pass, but is opaque to ultraviolet frequencies.*

Figure 6-7 shows a negative image made by exposing reversing film through positive artwork. Reversing film and developer are available from suppliers of pc board materials or photo supply shops.

The following process describes how to reverse artwork on a transparency using reversing film and developer.

Materials

You will need the following materials:

- Artwork to be reversed, on a transparency.
- Reversing film and developer.
- Ultraviolet lamp and mounting hardware. Possibilities include a 300-watt quartz halogen floodlight, #2 photoflood, 275-watt sunlamp, 15-watt fluorescent black light, or any light source that includes the range 350 to 425 nanometers. Even direct sunlight can be used. Special lamp stands are available or you can rig your own. Figure 6-8 shows an example of an ultraviolet lamp mounted on a stand.

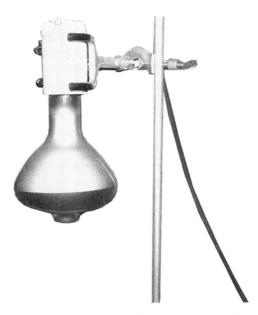

Fig. 6-8   *An ultraviolet sunlamp and mounting stand.*

- Contact frame or a flat surface and a pane of glass. Figure 6-9 shows a contact frame, also called a proofer. The contact frame consists of a box, typically lined with black foam, with a hinged glass cover. You place the artwork and

Fig. 6-9   *This contact frame, or proofer, holds the artwork flat against a hinged pane of glass for exposing.*

film to be exposed on the foam in the frame and close the cover over the artwork. The frame holds the artwork and film in close contact against the glass for exposing. If you don't have a contact frame, you can substitute a flat surface and a pane of glass to cover the artwork and film.
- Cotton balls for rubbing developer onto film.
- Rubber gloves.
- Scissors.
- Timer or clock with second hand.
- Lint-free gloves (optional). For film handling. These are available at photo supply shops.

Work area

Work in a room illuminated only by a yellow bug light, a lamp covered with a yellow filter, or very subdued incandescent lighting.

Do not look directly at the light from a sunlamp or other strong ultraviolet source. For maximum safety, operate the lamp with a remote timer.

For developing, you'll need a flat work surface and a source of running water for rinsing the board.

Procedure

When using reversing film or any photosensitive materials for the first time, you might want to make a test strip before you expose the artwork for the final project.

Making a test strip   When you use reversing film, many variables can affect the results, including the light's intensity, the distance between the light source and the image, and the exposing and developing time.

One way to find a combination that works well is to make a test strip. This is a strip of film on which you vary the exposure times on different areas. By first making a test strip and examining the results, you can find and document a procedure and timing that gives good-quality results without wasting a lot of time and materials on trial-and-error guesswork.

To expose a test strip, set up your lights and other materials as if you were exposing the artwork for a normal project. Instead of exposing the entire test strip for a specified time, make several exposures, including some for less than your estimated correct time, and some for more. After developing, pick the section of the test strip with the best results and use its timing for all of your projects.

The following paragraphs describe how to make a test strip for reversing film. Similar techniques can be used to make a test strip for any new process you try that has critical exposure times.

1. To make the test strip, you'll need a piece of artwork on a transparency at least ½ inch by 6 inches. The sample artwork should contain patterns similar to those in the artwork you'll be using in your projects.

2. Set up your equipment under normal lighting. Position your light source so the face of the bulb is 12 inches from the reversing film. Prepare a flat surface for the film in front of the light. Have your contact frame or glass handy to cover the film, a piece of cardboard or another opaque object for covering the test strip, scissors, and a timer, clock, or watch. Be sure the cover glass is clean, with no dust or dirt to block the light.

3. You're now ready to use the film. Turn on your subdued yellow lighting (if any) and turn off the room lights.

4. Carefully remove the reversing film from its lighttight bag. Handle the film only by its edges. Lint-free gloves will keep fingerprints off the film. Cut a piece of film slightly larger than the artwork to be reversed.

5. Identify the emulsion side of the film. This can be difficult to do in subdued light. The emulsion side is slightly duller, and the film tends to curl towards the emulsion side. Some film has a notch in the upper right corner of the emulsion side. If all else fails, you can scratch a corner of the film with an artist knife. If the orange coating scratches off, you have the emulsion side.

6. Place the film, emulsion side down, on the prepared flat surface or in the printing frame. Place the test artwork on top of the reversing film in its actual (right-reading) orientation so that a mirror image will be exposed onto the emulsion side of the film. Cover both with a pane of glass or close the contact frame to hold the artwork against the reversing film. Position the artwork so that it lies directly in the path of the lamp's beam.

7. Recommended distance and exposure time for a 300-watt quartz halogen floodlight or #2 photoflood is 2 minutes with the lamp 12 inches from the film. For the test strip, you might try exposures of 1, 1.5, 2, 3, 4, and 6 minutes. Turn on the lamp, start timing, and do the following at the times given:

- 1 minute: cover about 1 inch of the test strip with the cardboard.
- 1.5 minutes (30 seconds after the previous step): move the cardboard so it covers about 2 inches of the test strip.
- 2 minutes (30 seconds after the previous step): move the cardboard to cover a total of 3 inches of the test strip.
- 3 minutes (60 seconds after the previous step): cover 4 inches of the test strip.
- 4 minutes (60 seconds after the previous step): cover 5 inches of the test strip.
- 6 minutes (120 seconds after the previous step): turn off the lamp.

If you use a different light source or distance, adjust the exposure times, using longer exposures or shorter distances for weaker light sources, or longer distances and shorter exposures for more powerful sources or shorter distances. Sunlamps can get very hot, so don't place them too close to the artwork. If you want to try direct sunlight as a source, the estimated time is 30 seconds, but this will vary greatly depending on the location and conditions.

8. After exposing, you're ready to develop the test strip. In subdued light, remove the test strip from under the glass and place it, emulsion side up, on a smooth, clean work surface. Wearing protective gloves, pour a small amount of the reversing developer onto the film, and lightly rub with a cotton ball to spread the developer across the film.

   The developer causes the unexposed areas to wash away, leaving a clear image against an orange background. On a negative created from a positive image, the pads and traces will be clear. If nothing happens after about 30 seconds, you are developing the wrong side of the film. Flip it over and rub the developer on the other side.

9. When the image is visible and no more developer rubs off, turn on the lights, rinse the film thoroughly in running water, and hang it to dry.

10. When the test strip is dry, examine it to find the area with the highest contrast: transparent lines and pads on a colored background (assuming the original artwork was a positive image). The colored areas will be an orange that passes some visible light, but is opaque to ultraviolet frequencies. Figure 6-10 shows an example test strip on reversing film.

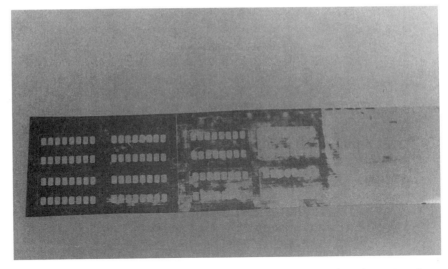

Fig. 6-10 *This test strip on reversing film shows the results of varying the exposure time.*

11. When you've decided on the best-quality image, calculate its exposure time from its position on the film. The end of the strip with the clearest background had the least exposure time (1 minute in our example), indicating that there was not enough light to harden the exposed areas. The opposite end (with a 6-minute exposure) might show color leaking onto the pads and traces, indicating that the long exposure time allowed light to penetrate the original positive image.

12. Make a record of the optimum exposure time along with the light source you used and how far away it was from the artwork. This is the setup you want to use when you use the reversing film on your board artwork.

Reversing the artwork   When you have your exposure time calculated, you're ready to reverse your pc board artwork. The materials and work area are the same as for making the test strip.

Under normal lighting, inspect the artwork and fill in any pinholes or breaks with opaque ink. Use a brush or squeezable bulb to remove dust from the image (Fig. 6-11). Don't neglect this step, because bits of dust or debris on positive artwork will show up as pinholes on your negative.

To expose the artwork, proceed as you did in making the test strip, cutting the reversing film about ½ inch larger than the artwork on all four sides. Use your calculated exposure time and

Fig. 6-11 *Use a brush or squeezable bulb to clean dust and dirt from the artwork before reversing it.*

develop, rinse, and dry as before. The result should be a reverse image of the original artwork: areas that were transparent are now colored, and areas that were opaque are now clear.

Small transparent flecks in the background are probably the result of dust or dirt on the original or on the glass cover. You can fill these in with opaque ink. If the background has transparent areas, try a longer exposure time. If the pads and traces contain colored areas that can't be explained by breaks in the original artwork, reduce the exposure time.

Step-by-step: Using pos-neg film

Pos-neg film from the Datak Corporation is a special film that, like reversing film, can convert positive or negative artwork on a transparency to the opposite form. Pos-neg film also has the unique ability to copy artwork on paper (a computer printout or magazine page, for example) onto a transparency. You can also use it to create a higher-contrast image from a weak one.

The pos-neg process uses a special high-contrast film that has unusual properties. When exposed to white light, such as the light from a photoflood, the film remains clear until it reaches a threshold, and then turns black all at once. If the exposed film is then exposed to a yellow light, it remains black until it reaches a threshold, and then quickly turns clear again. These properties make it possible to create positive and negative transparencies of pc board artwork. In addition to offering pos-neg materials, Datak publishes an instruction manual describing the theory behind the pos-neg process and how to use it.

The following section describes how to expose pos-neg film. After exposing, the film is developed like ordinary lithographic film. These developers are available from Datak or photographic suppliers. The developing procedure is described later in this chapter in the section titled "Developing Sheet Film."

Datak recommends making a test strip before converting your first artwork to find the exposure times that work best for your equipment and setup. To make a test strip, use the following procedures, but expose the strip in eight sections, for 15, 25, 35, and so on up to 85 seconds. Make a note of the exposure time that gives the best results. A slight brown haze on the transparent areas is acceptable. See the section earlier in this chapter for more on test strips.

Materials

You will need the following materials:

- Original artwork (positive or negative, on paper or transparency).
- Pos-neg film (available from Datak).
- Yellow filter (available from Datak).
- Contact frame or a pane of glass.
- 500-watt photoflood with color temperature of 3200°K to 3400°K (#2 photoflood, or ANSI type DXB, DXC, or DXE bulb) and suitable reflector and socket (available at photo supply shops).
- Lint-free gloves (optional). For film handling.
- Scissors.
- Timer or clock with second hand.

You'll also need the materials and equipment described in the section "Developing Sheet Film," later in this chapter.

Work area

You can work under subdued tungsten light (use an incandescent, not fluorescent, bulb), such as a 15-watt bulb that is 8 feet from your work area, or you can work in a photographic darkroom under a red safelight. Do not use a yellow bug lamp.

Procedure

The procedure for using pos-neg film varies depending on the type of image conversion you want.

Making a positive transparency from a positive on paper If your artwork is printed on paper, you can use pos-neg film to copy it to a transparency. Black areas will be opaque and white areas will become clear.

1. You should prepare and set out your developing chemicals in advance so they will be ready after exposing the sheet film. See "Developing Sheet Film" for this procedure.
2. Position the photoflood so that the front face of the bulb is 2 feet from the artwork's surface. For consistent results, build a box that holds the film and the lamp in fixed positions. Paint the inside of the box white for maximum light reflection.
3. In subdued light, carefully cut open the bag of pos-neg film. Handle the film only by its edges. Lint-free gloves will keep fingerprints off the film. Cut a piece of film slightly larger than your artwork. One side of the film is brown and shiny, the other is gray and dull. The dull, gray side is the emulsion side.
4. Place your positive artwork on your work surface or in the printing frame and cover it with the pos-neg film, with the brown, shiny side up. Cover the film with the yellow filter and close the contact frame or cover the filter with a pane of glass (see Fig. 6-12).
5. Turn on the photoflood and expose for 50 seconds or the recommended time from your test strip.
6. You're then ready to develop, fix, and wash the film. After doing these steps, the film will hold a copy of the original artwork, but on a transparency instead of paper. You can then use the transparency to expose your artwork onto a sensitized pc board, as described in chapter 7.

Reversing an image with pos-neg To make a negative from a positive on a transparency, expose the pos-neg film to yellow light, which clears the film. Then create a negative image on the film by exposing it through the artwork to white light.

As in the previous procedure, making a test strip is recommended to find the best exposure times for your equipment and setup. For this procedure, you should make two test strips: one with exposures from 20 to 90 seconds, for clearing the film, and another with exposures from 3 to 21 seconds, for reexposing the film.

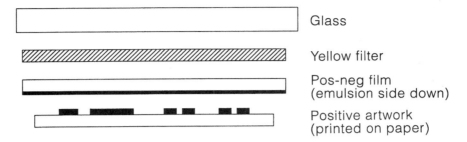

Glass

Yellow filter

Pos-neg film
(emulsion side down)

Positive artwork
(printed on paper)

Fig. 6-12 *Exposure setup for making a positive transparency from a
positive image on paper using pos-neg film.*

1. On your work surface or in your printing frame, place a sheet of pos-neg film cut to size with its gray, dull side up, covered by a yellow filter. Close the contact frame or cover the filter with a pane of glass. Expose for 55 seconds or the time you calculated with your test strip.
2. Replace the yellow filter with your positive artwork. Cover the artwork with glass or close the contact frame and expose for 12 seconds or the time calculated with your test strip (see Fig. 6-13).
3. Develop, stop, fix, and rinse. The result is a reverse of your original artwork: negative if the original was positive, and positive if the original was negative.

To improve the quality of a negative or positive, you can expose a piece of film through the original artwork and yellow light. More details on this are included in Datak's instruction booklet.

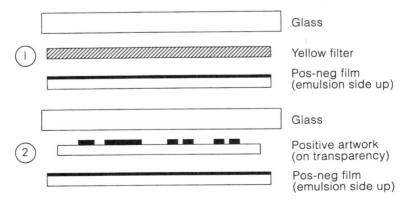

Glass

Yellow filter

Pos-neg film
(emulsion side up)

Glass

Positive artwork
(on transparency)

Pos-neg film
(emulsion side up)

Fig. 6-13 *Exposure setups for making a negative from a positive
(or a positive from a negative) using pos-neg film.*

Step-by-step:
Photographing pc board artwork

For high-quality results, you can prepare artwork at an enlarged scale on paper or on a transparency, photograph it, and then expose the resulting negative onto sheet film at actual size. When you develop the sheet film, you end up with a high-quality, actual-size transparency of the artwork, which you can use to expose a sensitized pc blank.

Camera photography gives excellent results if you have a critical board with fine traces and close spacing. You can prepare the original artwork at ×2 or even ×4 scale, and when the artwork is reduced to actual size, imperfections are reduced as well.

Camera photography is another way of transferring artwork on paper (such as a computer printout or magazine page) to a transparency for contact printing. It's also a good technique for creating a more-durable master artwork from a taping.

Photographing pc board artwork is similar to other black-and-white photography, with one main difference: pc board artwork consists of black and white images, with no intermediate shades of gray. To achieve finely detailed images that are pure black and white, a special high-contrast film is used. Otherwise the procedures and processes are similar to those used in ordinary photography.

If you don't want to do the photography yourself, a graphic arts printer will photograph your original artwork and reproduce it at actual size for you. If you want to try your own photography, the following sections describe how to photograph and develop pc board artwork using high-contrast film. If you want more information on photographing and developing, there are hundreds of books on photography and darkroom techniques. Check your local libraries and bookstores.

Materials

You will need the following materials:

- Artwork to be photographed, on paper or transparency, actual or enlarged scale.
- 35mm single lens reflex (SLR) camera with adjustable aperture settings (f-stops), capable of making timed exposures. An internal or external exposure meter is recommended for determining the correct exposure. You can use a standard 50mm camera lens, or another nondistorting

lens that allows the focused pc board artwork to fill the viewfinder.

- Tripod or copy stand for mounting the camera. Photographing pc board artwork requires exposure times as long as several seconds. During exposing, the camera must not move, shake, or vibrate, so a tripod or other stable camera mount is required. A handheld camera isn't steady enough.
- Locking cable release. This acts as an extension of the camera's shutter button, enabling you to shoot a picture without jarring the camera body. With a locking mechanism, you press the lever once to open the shutter, then again to close it, completing the exposure.
- 35mm high-contrast film, such as Kodalith ortho type 3 film (ASA 8). This type of film is also called orthographic, lithographic, or graphic arts film. This special-purpose film records high-resolution images in black (opaque) and white (clear) with no intermediate grays. The film is normally used in graphic arts for line-art photography of black-and-white drawings, and is perfect for pc board work. The film is sensitive to blue and green light frequencies and insensitive to red.
- Two or three 500-watt photofloods or a similar light source. For illuminating the artwork.

The film, photographic chemicals, and darkroom supplies are available at photographic supply shops.

Work area

To photograph the artwork, you need a place to set up your tripod or copy stand and lamps, and a flat horizontal or vertical surface for mounting the artwork to be photographed.

Procedure

The goal in photographing the artwork is to make a high-contrast negative of the artwork. The white or clear areas in the artwork will be opaque on the negative, and the black areas in the artwork will be clear.

1. Following the recommended procedures for your camera, load the camera with a roll of orthographic film.
2. Install a 50mm or other appropriate lens and the cable release on the camera.
3. Position the artwork and camera. You want the camera's

Fig. 6-14   *Setup for photographing pc board artwork with a 35mm camera.*

lens to point straight-on at the center of the artwork. The artwork must lie parallel to the lens, not tilted at an angle. If you are using a tripod camera mount, tape the artwork to a wall or other vertical surface with the camera lens aimed directly at it. Figure 6-14 shows an example setup. On a copy stand, the camera lens points down at the artwork lying beneath it. Position the camera so that the artwork nearly fills the field of view.

4. Position the lamps to illuminate the artwork fully and evenly. Aim the lamps at the image from two sides, from the top, from the back, or a combination. Adjust the lamps to eliminate glare on the artwork.

5. For best results, set the camera's aperture to a small opening, such as f/16 or f/22.

6. If possible, adjust the camera's shutter speed so that the light meter shows the proper exposure. For long exposures, you might have to control the shutter speed manually with the cable release.

7. Adjust the focus for a sharp image.

8. Take a photo.

9. To give a margin of error, take at least two additional photos, one with twice the exposure time and one with half the exposure time. For example, if the light meter

recommends a 2-second exposure, also take exposures of 1 and 4 seconds. This will help to ensure that one of the photos is correctly exposed. With practice, you might be able to reduce the number of exposures.

10. If you are making a double-sided board, you'll need to photograph two images, one for each side of the board. After you photograph the first artwork, don't move the camera. Just remove the first artwork, place the second one in the same position, and photograph the second image in the same way.

Film developing

After photographing the artwork, you're ready to develop the film. You develop orthographic film much like other black-and-white film, but using a special orthographic developer that gives the desired high-contrast results.

If you wish, you can send your film out for professional development. If you decide to do your own, the following sections describe the steps in setting up a darkroom and developing film.

Film developing basics

Photographic film has a coating of silver halide crystals suspended in gelatin. The suspension is commonly called the *emulsion.* When the emulsion is exposed to light, it undergoes a chemical change, and an invisible latent image forms on the film.

In developing, the film is immersed in a chemical bath, and the silver halide crystals that were exposed to light are reduced to metallic silver. This causes a negative image to form on the film, with areas that were exposed to light being dark and unexposed areas being clear.

Creating a darkroom

Some photographic procedures require a darkroom, which is an enclosed, lighttight workspace. You do not need a dedicated room for this; you can temporarily take over a bathroom, closet, kitchen, a corner of a basement, or another space.

A darkroom must be lightproof. Windows and other openings and cracks must be covered with opaque material. In addition, a darkroom must have the following:

- Surface areas for setting up developing trays, an enlarger, and other materials.

- Electrical outlets for a safelight and an enlarger, if used.
- Ventilation.
- Hot and cold water. A source of running water and a drain are handy, but not essential. You can bring in buckets of water as needed.

A bathroom is often the best choice for a temporary darkroom. Windows are usually few or nonexistent, ventilation fans and electrical outlets are often present, and running water is available. A bathtub or sink covered with plywood provides a work surface. But any room or space that meets the requirements can be used.

A darkroom is dark enough if, after 5 minutes inside, you see no cracks of light. To make sure a darkroom is dark, apply weatherstripping around door frames. If you work at night, closing the window shades might be sufficient. During daylight, tape a sheet of opaque material over the window, or make a removable panel that can be attached over the window.

Step-by-step:
Developing 35mm orthographic film

The following is the procedure for developing a roll of 35mm Kodalith type 3 orthographic film. Always check the instructions included with your film and developer for specific recommendations.

Materials

You will need the following materials:

- Roll of exposed 35mm Kodalith type 3 orthographic film.
- Kodak D-11 developer. Kodalith developer can also be used. If you are using a different film or developer, be sure to check and follow the instructions included. High-contrast film requires a specially formulated developer.
- Stop bath. This neutralizes the developer on the film and prevents it from contaminating the fixer. You can substitute a rinse in running water.
- Kodak fixer. This dissolves the undeveloped silver halide crystals on the film so they wash away.
- Developing tank. Inexpensive tanks are available that consist of a reel to hold the film and a lighttight, waterproof

Fig. 6-15   *Tank and reel for developing 35mm film.*

tank that holds the developing solutions and film reel (Fig. 6-15).

- Accurate, immersible thermometer. For checking developer temperature.
- At least two 32-ounce graduated beakers or measuring cups. Hand-labeled plastic cups are acceptable if you don't want to invest in special glassware.
- Funnel. For pouring chemicals without spilling.
- At least two dark glass bottles, or plastic bottles rated for use for storing photographic chemicals. The plastics used in ordinary bottles can react with the chemicals. Because exposure to oxygen decreases the shelf life of the chemicals, the size of the bottles should match their contents. You can squeeze plastic bottles to remove air, or add glass marbles to a partly filled bottle to fill any extra space.
- Timer or clock with a second hand.
- Protective gloves.
- Film clips or clothespins. For hanging film to dry.
- Scissors.

Work area

Always read and follow the safety precautions included with photographic chemicals and equipment. Have adequate ventilation, avoid contact with your skin, eyes, or clothing (wear rubber gloves), and wash your hands after handling chemicals. Kodak publishes a variety of booklets on health, safety, and environmental concerns relating to photography.

The roll of film is developed in an enclosed tank, so you do not need a darkroom for this step. However, when loading the film onto the developing reel and placing the reel in the tank, you must use a dark closet, a lighttight changing bag, or a darkroom illuminated by a red or amber safelight.

Procedure

Film developing consists of preparation, loading the film on the reel, developing, stopping development, fixing, and cleaning up.

1. Prepare the chemicals. Dry chemicals must be dissolved in water. Using the graduated beakers and the proportions given on the packages, prepare the solutions.
2. When you're ready to develop, you'll have more consistent, predictable results if you always develop film at the same temperature. For Kodak D-11 processing, 68°F is recommended. If you are working in a warmer or cooler room, place your bottles of developer in a sink or pan filled with water at the desired temperature (Fig. 6-16). Measure the water temperature periodically and add hot or cold water as needed. While the chemicals are warming or cooling, you can proceed with loading the film on the reel for developing.
3. Load the film on the reel following the recommended procedure for your reel. This step must be done in darkness, or illuminated only by a red or amber safelight. If this is your first experience loading film onto a reel, practice with a roll of undeveloped film until you feel confident about your abilities. Only then should you attempt to load your exposed film.
4. Insert the loaded reel into the developing tank, and close and seal the tank. You can now turn the lights back on.
5. Check the temperature of the chemicals to verify that they are at 68°F. If necessary, wait until the temperature is correct.
6. Consult the developer's data sheet for the recommended developing time. Make a note of it, or set your timer (but don't start timing yet).
7. Pour the required amount of developer, stop bath, and fixer into graduated beakers or labeled cups.

Fig. 6-16   *Place tightly capped bottles of chemicals in water to bring them to the desired temperature.*

8. Developing is the most critical step for timing. Try to move quickly and follow the recommended times. Quickly but carefully, remove the lighttight drain plug from the developing tank, start your timer or note the time, pour the developer into the tank, and replace the plug.

9. To dislodge air bubbles that might cling to the film after pouring in the developer, tap the tank sharply a few times against a hard surface. Immediately begin agitating the tank, as recommended by its manufacturer.

10. Stop development. When the developing time is almost up, begin pouring the developer out of the tank so that the tank is empty when the developing time is over. If the developer is reusable, pour it back into its bottle. Otherwise, pour it into a container for disposal.

11. When the tank is empty, quickly pour in the measured amount of stop bath and agitate for 15 seconds. If you are using water as a stop bath, pour out the first bath after 15 seconds and repeat.

12. When the stop bath time is up, pour the stop bath out of the tank. If the stop bath is reusable, pour it back into its bottle. Otherwise, pour it into a container for disposal.

13. Fix the negative. Pour the fixer into the tank and agitate for the recommended time. When the fixing time is up, pour the fixer back into its bottle for reuse.

14. Open the developing tank and rinse the reel and film in running water that is 70°F to 75°F. To be sure all traces of chemicals are removed from the film, fill and dump the tank 20 times or more.

15. Dry the film. Hang the film to dry in a place where it will be undisturbed and that is dust-free. You can hang the film on a line with one clothespin on top to hold the film on the line, and another on the bottom to prevent the film from curling. Drying takes several hours, unless you speed it up with a blow dryer.

16. Clean up. Remember to rinse all of your equipment thoroughly after use to remove chemical residues.

Step-by-step:
Enlarging a 35mm negative

When you have your developed negative, you're ready to enlarge it onto sheet film, which you then develop to obtain a positive, actual-size image of your artwork on a transparency.

Materials

You will need the following materials:

- Developed 35mm negative of artwork. This is the image you developed in the previous step. Handle the negative only by its edges. If you took several bracketed shots as recommended, choose an image with good contrast, with opaque and transparent areas only, as shown in Fig. 6-17.
- Orthographic sheet film, such as Kodalith ortho type 3 film (ASA 8). This is the same film used in the camera, but in 4-by-5-inch or 8-by-10-inch sheets instead of 35mm size, for holding the enlarged (actual-size) artwork. Keep the film in its lighttight box until you are ready to use it.

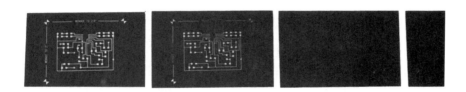

Fig. 6-17 *Photograph the artwork at a variety of exposure times and select the negative with the best contrast for enlarging.*

Fig. 6-18 *Photographic enlarger for exposing 35mm artwork onto sheet film.*

- Photographic enlarger (Fig. 6-18). The enlarger projects an image onto a flat surface and functions much like a slide projector, but with a precise, distortion-free image. To vary the size of the projected image, adjust the height of the film carrier that holds the negative. An aperture adjustment enables you to control the size of the lens opening, as on a camera.

In pc board work, the enlarger projects the roll film's negative image onto sheet film to expose a positive, actual-size transparency of the artwork. The enlarger is the most expensive piece of equipment required, but it's useful for conventional photographic work as well as in pc board work.

- Timer or clock with a second hand.
- Ruler. For measuring the enlarged image.

Work area

To enlarge sheet film, you need a photographic darkroom with a 15-watt red or amber safelight mounted at least 4 feet from the film. You can use a red- or amber-coated safelight bulb or a dome safelight with a type 1A red filter.

Procedure

You should prepare and set out your developing chemicals in advance so they will be ready after exposing the sheet film. See "Developing Sheet Film" for this procedure. If you're not sure of the exposure timing, expose and develop a test strip at a variety of exposure times before exposing the artwork for a board.

1. Prepare the negative. Be sure it is dust-free. Use compressed air or a squeezable rubber bulb to blow away dust or dirt on the negative.
2. Place the negative in the enlarger's film carrier, emulsion side down, so that the projected image is oriented the same as the original artwork. (With the negative's emulsion side up, the projected image will be a mirror image of the original artwork.)
3. Turn on the enlarger.
4. Open the lens aperture to maximum.
5. Focus the projected image on the enlarger's baseboard.
6. Adjust the height and focus until you have a sharp, projected image of the desired dimensions. Measure the dimensions carefully, because an image that is even slightly smaller or larger than the intended size will make it hard to install components and align double-sided boards.

   If your artwork includes targets, you can use these to measure. Otherwise, you can measure the distance between IC pads in the artwork. Be sure to measure along both dimensions of the image to ensure that the negative is parallel to the projected image. If the original artwork was drawn actual size, you can place it on the enlarger's baseboard, and compare its size and scaling to the projected image.

8. Set the aperture and timer for proper exposure, typically about 10 seconds at an aperture of f/16. There is a wide latitude for effective exposure times, so precise timing isn't required.

9. Turn off the enlarger and room lights, except for the safelight. Place a piece of sheet film on the enlarger's baseboard.

   For best results, when you expose the pc board during final image transfer, you want a mirror image of the actual pc board artwork exposed onto the emulsion side of the ortho film. Doing so ensures that the emulsion side of the film contacts the pc board during image transfer.

   A notch might indicate the emulsion side of the film. When the notch is in the upper right corner of the film, the emulsion side is up. Also, the emulsion side is usually duller.

   If your pc board uses positive-acting photoresist, and if your projected artwork shows the actual orientation as it will appear on the pc board, place the sheet film emulsion side down. If the projected artwork is a mirror image of the artwork as it will appear on the pc board, place the sheet film emulsion side up.

   If you'll be reversing the sheet film to a negative image, you want to expose a mirror image of the actual orientation onto the emulsion side. Then when you expose the positive image onto film, you place the films emulsion-to-emulsion to expose the actual image onto the emulsion side of the negative.

   All of this reversing and mirror-imaging can get confusing. If you make a mistake and expose the wrong side of the film, don't panic, the artwork is still usable. On boards with very fine traces, the image quality might be slightly lower if you expose the pc board with the film emulsion side up.

10. Expose the film for the recommended time.

Step-by-step: Developing sheet film

Developing sheet film is similar to developing roll film, except that the film is placed in trays instead of in a tank. (Tanks for sheet film are available for processing large quantities of sheet film, but only the simpler, cheaper tray method is covered here.) Developing sheet film involves preparation, developing, stopping development, fixing, and cleaning up. The same procedure is used whether you are developing a positive image, a negative image prepared from a positive, or the special pos-neg film described earlier.

Materials

You will need the following materials:

- Exposed sheet film, orthographic or pos-neg.
- 3 plastic or glass developing trays, 1 inch deep and large enough to hold sheet film the size of your artwork (Fig. 6-19).
- Bamboo or stainless steel tongs.
- Immersible thermometer.
- Graduated beakers or other calibrated cups. For measuring developing chemicals.
- Funnel. For pouring chemicals.
- Chemicals: developer, stop, and fixer recommended for your film.
- Protective gloves.
- Timer or clock with a second hand.

Fig. 6-19 Trays for developing sheet film.

Work area

To develop sheet film, you need a photographic darkroom with a 15-watt red or amber safelight mounted at least 4 feet from the film (the same lighting as required for enlarging). You can use a red- or amber-coated safelight bulb or a dome safelight with a type 1A red filter.

Wear rubber gloves, have adequate ventilation, and follow all recommended safety procedures for your chemicals.

Procedure

Follow these simple steps:

1. Prepare the chemicals according to their instructions. As before, bring the chemicals to 68°F.
2. Set out three trays and pour the recommended amounts of developer, stop bath, and fixer into each. To prevent contamination, use a fresh measuring cup for each chemical, or rinse the cup thoroughly after each use. Set a pair of tongs in front of each tray, or use one set of tongs and prepare a beaker of water to rinse the tongs after each use.
3. Be sure your timer and film are handy.
4. Develop the film. Start the timer and immediately place the exposed film in the developing tray. Agitate by gently rocking the tray.
5. Stop development. When the development time is up, remove the film from the developer, place it in the stop bath, and agitate as before for about 30 seconds.
6. Fix. Transfer the film to the fixer and agitate. After a couple of minutes in the fixer, you can turn on the lights because the film is no longer light-sensitive.
7. Washing. After the stop bath, thoroughly rinse the film in running water that is 70°F to 75°F.
8. Dry the film. Use a clothespin or film hanger to hang the film to dry in a place where it will be undisturbed and that is dust-free.
9. Clean up. Remember to rinse all of your equipment thoroughly after use to remove chemical residues.

Step-by-step: Reversing an image on orthographic film

After developing the sheet film, you will have a positive transparency of your artwork (opaque pads and traces, clear back-

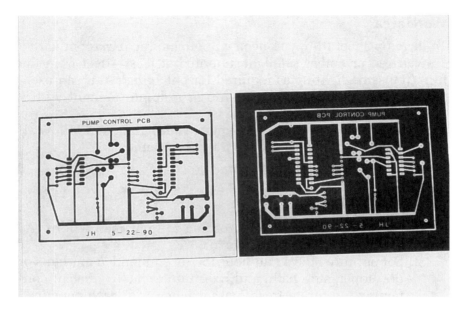

Fig. 6-20 *Positive and negative artwork.*

ground), assuming that the original artwork was a positive. If you are using pc boards with positive-acting sensitizer, you can use your positive transparency to expose the pc board. If you are using a negative-acting sensitizer, you have an additional step remaining: creating a negative image from your positive so that you have clear pads and traces on an opaque background.

You can reverse the transparency using the reversing film or pos-neg processes described in chapter 5, or you can contact print and develop another sheet of high-contrast film. In contact printing, the artwork on a transparency is placed on a fresh piece of sheet film, and the film is exposed. After developing, the result is a negative image of the original artwork. Figure 6-20 shows an example of positive artwork and a negative transparency that was made by contact printing.

Materials

You will need the following materials:

- Actual-size transparency of artwork to be reversed.
- Sheet of orthographic film.
- Contact frame or a pane of glass.
- Photographic enlarger.

Work area

The work area for contact printing onto orthographic film is the same as that for printing with an enlarger: a darkroom illuminated by a red or amber safelight.

Procedure

Follow these steps:

1. Place the artwork to be reversed in the contact frame or directly on the enlarger's baseboard. The artwork should show a mirror image of the artwork as it will appear on the pc board.
2. Turn on the enlarger and adjust it so that the light covers the entire artwork. Adjust the focus for sharp edges on the projected image.
3. Set the enlarger's aperture to the same setting used for enlarging the 35mm negative. The exposure time is the same as well.
4. Turn off the enlarger.
5. Working under a safelight, remove a piece of sheet film from its lighttight box. Place the film, emulsion side up, in the contact frame or directly on the enlarger's baseboard, with the artwork to be reversed on top of the film. Orient the artwork so that a mirror image of the artwork as it appears on the pc board is exposed onto the emulsion side of the negative.
6. Close the contact frame or cover the film with a pane of glass to hold the two layers in tight contact.
7. Turn on the enlarger and expose the film for the same amount of time you used to expose the original sheet film.
8. Develop in the same way you developed the original artwork on sheet film.

7
Transferring the artwork to the pc board

When your pc board artwork is designed and in the proper format, you're ready to transfer the image onto the board. The goal is to create an etch-resistant, actual-size image of the artwork on the pc board. In the etching step that follows image transfer, the exposed copper is removed and only the artwork remains drawn in copper on an otherwise bare board.

Many different methods exist for transferring artwork to a board, including some new, easy-to-use, and very effective ones. You might want to experiment with a variety of methods to find the one that works best for you and your circuits. The image transfer methods discussed in this chapter include

- Transfer patterns—you lay the pads and traces directly on the board for etching.
- Iron-on transfers—you laminate a dry-toner image from a photocopier or laser printer onto a copper-coated board.
- Direct plot—a pen plotter draws the artwork in etch-resistant ink directly onto copper.
- Contact printing—you lay a transparency containing the artwork on a presensitized board which you then expose and develop.
- Screen printing—the artwork's image is transferred to a fine screen that serves as a stencil for printing resist ink onto the pc board.

Also included in this chapter are techniques for registering the top and bottom images when making double-sided boards.

Step-by-step: Using transfer patterns

For simple, one-of-a-kind circuits, you can place transfer patterns directly on a copper-coated board. This is a quick way to transfer artwork for simple circuits. If you have a circuit with more than a few components, placing the patterns accurately and neatly can be time-consuming and difficult to do well. If you need multiple copies of a circuit, other methods will do the job more quickly.

Materials

You will need the following materials:

- The pc board artwork drawn to scale. For a simple circuit, the artwork doesn't have to be precise, but be sure you have at least a sketch to work from.
- Clean, copper-coated pc blank.
- Selection of press-on or rub-on transfer patterns and tapes (see chapter 5). Whether to use press-ons or rub-ons, or a combination, is up to you. Do not use press-on IC patterns that have a transparent carrier film that transfers along with multiple pads, because the carrier film prevents the etchant from reaching between the pads.
- Burnisher. The back of a spoon or comb will do for this.
- Resist pen. Most permanent marking pens will function as an etch resist. For touch-ups.
- Carbon paper (optional). For rough image transfer before laying down the patterns.

Work area

You'll need a flat work surface with good lighting to transfer the artwork to the board.

Procedure

Follow these simple steps:

1. Clean the board thoroughly, as described in chapter 6.
2. For a simple circuit, the pads and traces can be positioned on the board by eye. As an alternative, lightly sketch component outlines in pencil on the board, or lay a piece of carbon paper on the board, cover it with your artwork, and trace a sketch of the artwork onto the board. You don't have to copy every detail onto the board, but include enough information so that you can place your pads and traces accurately.
3. When you are ready to lay the pads and traces, use the techniques described in chapter 5 for making original art-

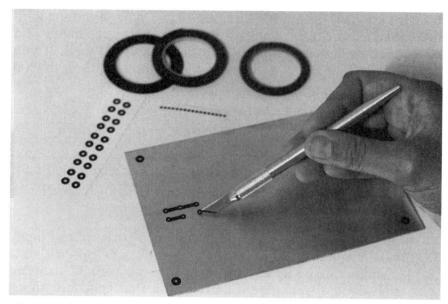

Fig. 7-1 *Laying transfer patterns directly on a copper-coated board.*

work with press-on or rub-on patterns, but place the patterns directly on the board instead of on a transparency
(Fig. 7-1). Try not to touch the copper with your bare
hands as you work.

4. When you're finished, check your work carefully and correct any mistakes or flaws. You can remove patterns by
lifting or lightly scraping with an artist knife. When the
artwork is correct and complete, burnish, and you're
ready to etch.

Step-by-step: Iron-on image transfer

Iron-on transfer is a quick, simple, and low-cost way to transfer
artwork to a pc board. In this method, you use a photocopier or
laser printer to print the artwork onto a special film or paper,
then transfer the artwork onto a bare copper pc board using a
clothing iron or other heat source and moderate pressure.

There are several similar products suitable for iron-on image
transfer, including coated paper from DynaArt Designs, blue-
coated Press-n-Peel film from Techniks, and transparent TEC-200
film from Meadowlake. Figure 7-2 shows examples of artwork
printed on each of these.

Iron-on transfer gives good results with traces as narrow as
0.015 inch, spaced 0.015 inch apart. With a little practice, you

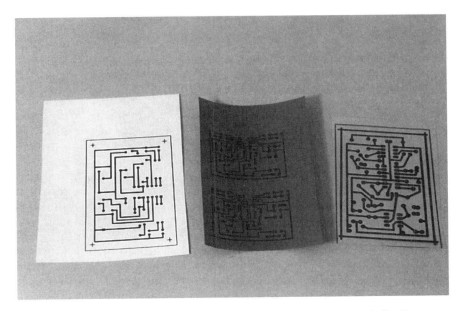

Fig. 7-2   *The pc board artwork printed onto DynaArt paper (left), Press-n-Peel film (center), and TEC-200 film (right).*

can develop techniques that work reliably and consistently. At about $1 per sheet, you can try all three products and see which you prefer. The processes are described below.

Using DynaArt's toner transfer paper

In the DynaArt method of image transfer, you laser print or photocopy your artwork onto a special coated paper. To transfer the artwork to the pc board, place the paper, image side down, on a copper-coated board and apply heat and pressure with a clothing iron or other heat source. Immerse the board and paper in water. The water-soluble coating dissolves and the paper lifts from the board, leaving the artwork's image fused on the board in the etch-resistant toner. Figure 7-3 shows the materials you'll need.

Materials   You will need the following materials:

- The pc board artwork, mirror-image orientation, actual size, for photocopying or in an appropriate format for computer printing.
- Copper-coated pc blank.
- DynaArt toner transfer paper.
- Heat-resistant cover sheet (included with the transfer paper).

Fig. 7-3 *Materials for transferring artwork using DynaArt iron-on film.*

- Photocopier or laser printer.
- Clothing iron or other heat source (about 300°F).
- Bowl, bucket, tray, or sink of water large enough to immerse the pc board.
- Tongs or other tool for lifting the hot board and placing it in the water bath.
- Resist pen or permanent marker. For touch-ups.

Work area   You'll need a work table or another heat-resistant, flat surface for ironing the image.

Procedure   The image transfer has two main steps: copying or printing the image on the paper, and transferring the image from the paper to the board.

1. To copy or print onto the transfer paper, use any copier or printer that forms images by fusing a dry toner onto paper or film. This includes most laser printers and photocopiers.

    Before you copy onto the transfer paper, make a test copy or printout on ordinary paper. You want to print or copy a mirror image of the artwork's orientation as it will appear on the board. In other words, for a single-sided, through-hole board, you want to print the top, or component-side orientation, while for a single-sided, surface-mount board, you want to print a mirror image of the component-side orientation. The images mirror back to their correct orientations when they transfer to the board.

If necessary, you can create a mirror image with a photocopier. Photocopy the artwork onto a transparency, flip the transparency over, and copy this image onto a fresh transparency or sheet of paper.

2. Verify that your test image is scaled correctly and has the correct orientation. Examine the artwork for breaks or pinholes. Tiny pinholes often fill in by themselves as the toner melts during image transfer. If you are photocopying the artwork, fill voids with a resist pen or permanent marker before you copy. The better your original artwork is, the better your copy will turn out. Chapter 6 has other tips on how to get a good-quality image on a photocopier or laser printer.

3. When you have a good-quality test image, you're ready to print or photocopy the artwork onto the transfer paper. The sheets can shrink or stretch slightly if exposed to high humidity, so store them in their package until you're ready to use them.

When you remove the transfer paper from its package, handle it by its edges to prevent contamination by finger oils or dirt. The coated side, which is the one you want to print on or copy to, is shinier than the uncoated side.

If you're printing from a computer file, load the transfer paper into the printer's paper bin so that the printout appears on the shiny side. If your printer has a choice of paper feeds, use the one recommended for printing onto transparencies.

4. Examine the resulting image. Look for opaque, black pads and traces against a white background. If the image is flawed, try again, adjusting the contrast as necessary.

5. When you have a good-quality image, you are ready to transfer it to a pc board. Be sure that the copper surface of the board is flat, without burrs or raised edges that will prevent the iron from applying even pressure to the entire board. If necessary, file the edges flat.

6. Clean the pc blank, following the recommendations in chapter 6. A clean board is essential for good image transfer. After cleaning, let the board dry thoroughly. Because the transfer paper has a water-soluble coating, you don't

want it to contact water until after the image is transferred.

7. While the board is drying, get ready to transfer the artwork to the pc board. Prepare a solid, heat-resistant surface, such as a workbench or an old table.

8. Plug in and turn on the clothing iron. Set the temperature to 300°F (cotton setting). Fill a bowl, bucket, or sink with water to immerse the board in. Cut the artwork from the DynaArt sheet in a piece about the same size as the pc board.

9. Carefully position the artwork, toner side down, on the clean copper surface of the pc blank (Fig. 7-4). Place the heat-resistant cover sheet over the transfer paper to provide a smooth surface for the iron.

10. Note the time. Place the iron on the pattern and begin moving the iron in circular motions over the image (Fig. 7-5). Use light pressure, applying the weight of the iron and your arm. Cover the entire image area evenly. Be especially sure to include the corners and edges.

The amount of time to iron varies with the size of the board and doesn't have to be calculated precisely. For small boards, iron for 1.5 to 2 minutes. Larger boards might require 4 to 5 minutes. In general, excess ironing time (up to 8 to 10 minutes) will do no damage, but too

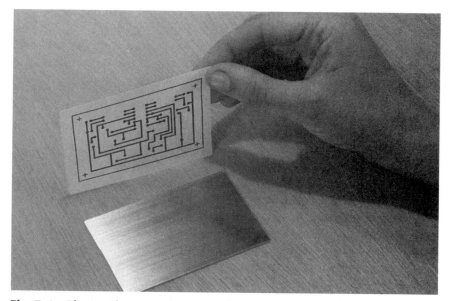

Fig. 7-4 *Placing the artwork on the clean, copper-coated pc board.*

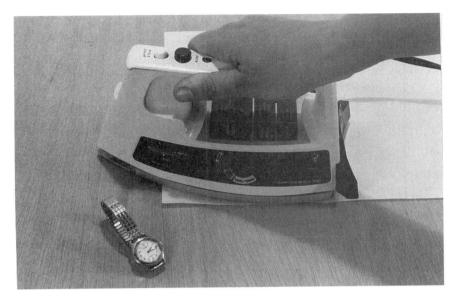

Fig. 7-5   *The heat and pressure of the iron transfers the artwork onto the pc board.*

little time can result in areas that don't adhere to the board. So don't skimp on ironing time, and try to give equal time to all areas of the image.

11. When the ironing is complete, you're ready to remove the transfer paper. Carefully lift off the cover sheet. Use tongs or a towel to lift the hot board, with the transfer sheet attached, and place it in the prepared water bath (Fig. 7- 6). After a minute or so, the coating will dissolve and the paper will float free from the board.

12. Handling the board by its edges, remove it from the water and shake off excess water (Fig. 7-7). Inspect the pattern. You should see a complete, good-quality image of your artwork on the board, with solid pads and traces and no missing or smeared areas.

If all looks okay, the board is ready to etch. You can fill in small imperfections with a resist pen or permanent marker (Fig. 7-8). If the transfer was unsuccessful, try again. It might take some practice to develop a technique that works for you. These are some possible problems and likely causes:

- Image did not adhere to the board—insufficient heat or pressure. Increase ironing time, temperature, or pressure.

Fig. 7-6   *Placing the board with its transferred artwork into a water bath.*

Fig. 7-7   *Removing the board from the water bath.*

- Good quality image, but some areas did not transfer (Fig. 7-9)—uneven ironing. Concentrate on giving equal attention to all areas, especially edges and corners.
- Smeared or distorted image (Fig. 7-10)—too much heat or pressure. Reduce ironing time, temperature, or pressure.

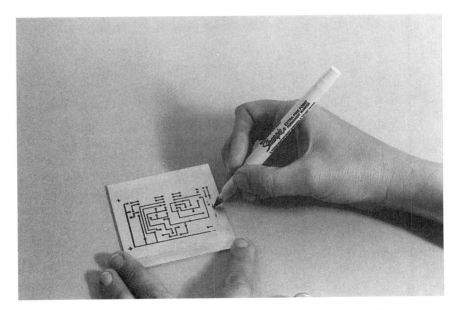

Fig. 7-8   *You can use a resist pen or permanent marker to touch up areas of artwork that didn't transfer.*

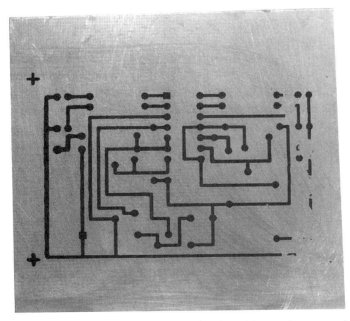

Fig. 7-9   *Artwork that didn't transfer completely. This can be caused by too little heat, pressure, or time.*

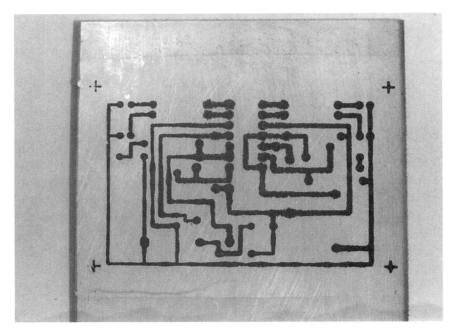

Fig. 7-10 *Smeared, distorted image, caused by too much heat, pressure, or time.*

Using Press-n-Peel film

Press-n-Peel is a transparent image transfer film with a dark blue coating. You use Press-n-Peel much like the DynaArt film described previously, with these differences:

- Print on the dull, matte side of the film, not the shiny side.
- Recommended temperature for ironing is 200°F to 225°F (lower region of steam setting).
- Iron for 30 to 60 seconds. When the image has transferred, it appears more pronounced through the film.
- After ironing, do not immerse the board in water as you do with DynaArt film. Instead, leave the board to cool. When it has cooled, begin at one corner and carefully lift the Press-n-Peel film from the board (Fig. 7-11). The artwork will remain fused to the board. The blue coating fuses with the toner and transfers along with it, so the removed Press-n-Peel film shows a negative image of the transferred artwork (Fig. 7-12).
- Inspect and touch up as usual.

Fig. 7-11 *Lifting Press-n-Peel film from a board.*

Fig. 7-12 *The removed Press-n-Peel film holds a negative image of the transferred artwork.*

Using TEC-200 film

Meadowlake Corporation's TEC-200 is another transparent film suitable for iron-on transfer. You can also experiment with other transparencies made for use in photocopiers or laser printers. Some brands work better than others. If you experiment, be sure the transparencies you use are suitable for photocopier or laser printer use. If the label on the box doesn't recommend these uses, do not use the film, because some transparencies are not heat resistant enough to withstand the heat of a laser printer or photocopier. Using the wrong type of transparency can damage or ruin a photocopier or laser printer.

You use TEC-200 film much like the DynaArt film described previously, with these differences:

- Print on either side of the film. Because the film is transparent, be sure to keep track of which side you printed on or photocopied to. If in doubt, use an artist knife to scratch a noncritical portion of the image. The image will scratch away on the toner side.
- Recommended temperature for ironing is 265°F to 295°F (cotton/linen setting).
- As with the Press-n-Peel film, after ironing, set the board aside to cool. When the board is cooled, begin at one corner and carefully lift the film from the board (Fig. 7-13).
- Inspect and touch up as usual.

Fig. 7-13 Lifting TEC-200 film from a board.

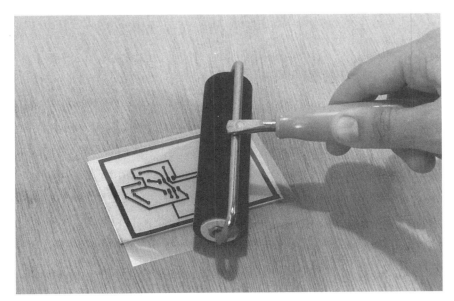

Fig. 7-14 *Using a rubber print roller to transfer the artwork onto a preheated board.*

Step-by-step:
Alternate iron-on transfer method

Meadowlake also recommends an alternate method for TEC-200 image transfer that often works better for smaller circuits. The following paragraphs describe this method.

Materials

This method requires the same materials as any iron-on image transfer, plus a handheld rubber print roller, or brayer, which is available in photo supply shops.

Procedure

Follow these simple steps:

1. Print or photocopy the artwork onto the TEC-200 film as usual.
2. Prepare your work area and plug in and turn on the clothing iron, setting it to its cotton/linen setting.
3. To transfer the image, heat the board, then quickly place the artwork, toner side down, on the board and transfer the image onto the board with the rubber roller. Because

you get only one chance to position the artwork on the board, practice with an unheated board first.

4. When you're ready, lay the hot iron on the bare copper board. If the plate of the iron isn't large enough to cover the board's surface, move the iron around so that all areas are heated equally.

5. When the board is heated (30 to 60 seconds), remove the iron and quickly and carefully lay the artwork, toner side down, on the heated copper surface. Move the rubber roller across the transparency, using moderate pressure and varying the direction (Fig. 7-14). Roll across the image several times.

6. After rolling the image onto the board, let the board cool. When it has cooled, carefully peel the transparency from the board and examine the transferred image. If a few breaks or pinholes remain, touch them up with a resist pen.

Step-by-step:
Direct plotting onto copper

In the direct plot method of image transfer, the artwork is plotted directly from a computer onto a thin sheet of copper-coated base material using a pen plotter and etch-resistant ink. General Consulting is the developer of this method, which they call Fast-Proto, and they sell most of the materials needed and a complete instruction manual (Fig. 7-15).

Materials

You will need the following materials:

- Pen plotter. This is the most expensive item required. Unlike printers, which use dot matrix, dry toner, or other technologies, pen plotters draw an image by moving a pen across a sheet of paper, much as you would if drawing by hand. Traditionally pen plotters have been used for making detailed, precise technical drawings.

  There are several reasons why pen plotters are feasible for plotting etch resist onto copper: the plotter bed can grip and feed the stiff copper sheets, refillable pens are available for the resist ink, and the plotters are precise and accurate enough to draw the high-resolution artwork required. Plotters you can use include Hewlett Packard's 7400 and 7500 series, Houston Instruments plotters, and others.

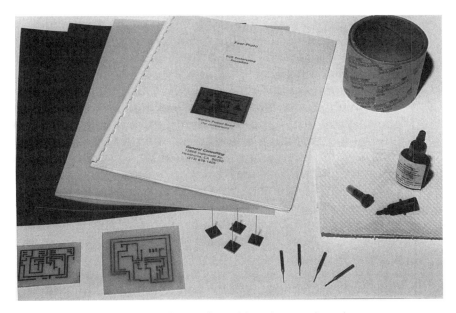

Fig. 7-15 *Materials for making a board by plotting directly onto copper.*

For highest resolution, General Consulting recommends using a servo-driven plotter. These can plot traces as narrow as 0.01 inch, with 0.01-inch spacing. Servo plotters are also more likely to have adjustable pen pressure, which gives longer pen life. But even the less expensive stepper motor plotters can plot traces down to 0.02 inch wide, with 0.02-inch spacing.

Unfortunately, pen plotters are expensive. Even basic desktop models cost $1000 or more, so it's hard to justify buying a plotter just for board-making. But if you already have a plotter or can borrow one, you're set. You might also be able to find a good deal on a used plotter.

- Computer and CAD software. The manual provided by General Consulting includes specific instructions for using PCAD, OrCad, Schema, Pads 2000, and AutoCAD software, but you can use any software that draws an accurate, actual-scale image of the artwork with solid traces and pads. Of course, the software must also include a driver for your plotter.
- Copper sheets—8½ by 11 inches. These have a copper coating on one side of a 0.005-inch thick FR-4 base. A very smooth copper surface and antioxidant coating help to ensure that the resist ink adheres. The copper thickness is

2 ounces per square foot (0.0027 inch), which is twice as thick as the 1-ounce copper used on many boards. This means that you might use up your etchant faster, and etching might take longer.

- Base material—FR-4 base material, 0.031 inch thick, 8½-by-11-inch sheets. Using this base material and the copper sheets provided, a finished double-sided board is about 0.041 inch thick. This is thinner than a typical ⅟₁₆ -inch (0.063-inch) board, but should cause no problems with most circuits. Boards with transformers or other heavy components might require a thicker base, which you can make by laminating two pieces of base material together.
- Plotter pen. In the refillable pens used for plotting with the resist ink, the pen tip is a fragile, thin wire that protrudes slightly from the body of the pen. The wires wear out after 6 to 10 square feet of plotting, and the pens are expensive, but you can save money by exchanging your worn-out pens for refurbished ones.
- Plotter pen adapter. The pen adapter allows the pen to fit in the plotter's pen holder. Different adapters are required for different plotters.
- Etch-resist ink. The ink is another expensive item, but an ounce goes a long way.
- Adhesive. General Consulting uses 3M's #9469 high-strength, silicon-base contact adhesive for laminating the copper to the base material. The adhesive comes on 3-inch wide rolls with a backing paper on one side. The adhesive can withstand soldering temperatures and solvents.
- Acetone. For pen and ink cleaning.
- Ultrasonic cleaner (optional). For pen cleaning. You can clean the pen by soaking it in acetone, but an ultrasonic cleaner is convenient and might be more effective.
- Handheld rubber roller. Available at photo supply shops.
- Tin snips. For cutting the base material.
- Scissors.

Procedure

Follow these simple steps:

1. Prepare the artwork using a computer and the CAD software of your choice.
2. Prepare to plot. Be sure you are plotting the correct orientation of the artwork. Because you are plotting directly

onto copper, plot the actual image of the artwork as it will appear on the board. If your software draws only a mirror image, General Consulting has a free program that mirrors plotter files to the correct orientation.

3. Select a plotting speed of 2 inches per second or less and a line width of 0.007 inch. Slow plotting allows the ink to flow more evenly on the copper. You can select pen speed and width through your plotting software or by programming the plotter directly.

4. Make a test plot on paper or polyester film, using a standard plotter pen (not the Fast-Proto pen). When you're satisfied with the test plot, you're ready to plot a board.

5. Handling the copper sheet by its edges or bottom, cut a piece about 1 inch longer and wider than your artwork and tape it to a sheet of Mylar. Scissors will cut the copper. Use your test artwork as a guide in positioning the copper on the Mylar so the plotted image is centered on the copper. Removable tape holds the copper well enough and is easy to peel off.

6. Pour some resist ink into the pen's ink reservoir (you don't have to fill the reservoir full for small jobs) and test the pen by drawing by hand on a scrap of copper. The pen should draw a thin, even line. If it doesn't, see the section following on pen care. Attach the pen adapter and install the pen in the plotter.

7. Plot the artwork (Fig. 7-16). At 2 inches per second, this takes a while.

8. Inspect the plotted copper. As needed, use the pen or an artist knife to touch up the artwork. If you make a mistake in plotting, you can clean the copper with acetone and try again, although the ink might not adhere as well the second time around. Use a fresh sheet for critical plots.

   After plotting, your artwork is ready for etching using the techniques described in chapter 8. General Consulting recommends using only ferric chloride etchant.

9. After etching, use the contact adhesive to laminate the artwork to a piece of base material. Use tin snips to cut a piece of the base material slightly larger than the size of your etched artwork.

10. Cover one side of the base material with the pressure-sensitive adhesive tape, and use a rubber roller or similar tool to press the adhesive onto the base material. For a double-sided board, repeat this step on the other side (Fig. 7-17).

Fig. 7-16 *Using a pen plotter to plot the artwork in resist ink directly onto copper.*

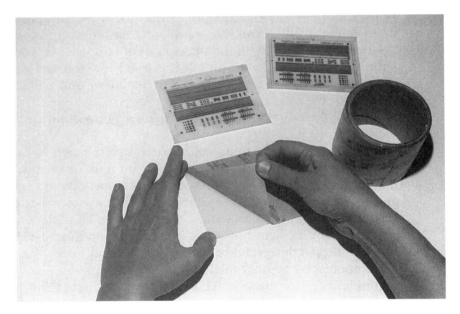

Fig. 7-17 *Lifting the backing paper from the pressure-sensitive adhesive applied to the base material.*

11. On one side of the base material, remove the adhesive's backing paper and carefully press the etched copper, image side up, onto the exposed adhesive. Press with the

roller to ensure good adhesion. For aligning the layers of double-sided boards, see the section "Image transfer for double-sided boards" later in this chapter.

Pen care

If you're not going to use a pen for 48 hours or more, clean it with acetone after use. You can remove minor clogs by wiping the pen tip across a piece of soft tissue. Pouring a little acetone on the tissue helps to dissolve small clogs, and briefly soaking the pen in acetone before use also helps to keep the pen from clogging.

Contact printing onto a pc blank

In contact printing, light and chemistry are used to transfer an image to a pc blank. The pc blank must be coated with a photosensitive material, which you can apply yourself, or you can buy precoated boards. To transfer the image, place a transparency containing the artwork on the coated board and expose the board to light. After exposing and immersing the board in a chemical developer, the board holds an etch-resistant image of the artwork, and you're ready to etch.

You can choose among several types of materials for contact printing, including positive-acting or negative-acting photoresists, solvent-based or aqueous developers, and precoated or hand-applied sensitizer. The following sections explain the options.

Positive-acting and negative-acting photoresists

There are two types of photoresist for use in two types of photographic image transfer: positive-acting and negative-acting. Positive-acting photoresist becomes soluble when exposed to light and requires positive artwork with opaque traces and pads on a transparent background. With this method you place the artwork on a board coated with positive-acting sensitizer and expose the board to a light source. When the board is immersed in its developer, the areas that were exposed to light (the background) dissolve away, and the unexposed areas (the pads and traces) remain as etch resist.

In contrast, negative-acting photoresist hardens when exposed to light. Negative-acting photoresist requires negative artwork with transparent pads and traces on an opaque background. The exposing and developing processes are similar.

Place the artwork on the coated board, expose it to light, and immerse it in developer. This time, the exposed areas (the pads and traces) remain as the etch resist, while the unexposed areas (the background) dissolve away. The result is the same, with the artwork remaining as etch resist on an otherwise bare board.

Most artwork is drawn as a positive image, whether it's laid out with transfer patterns, printed by computer, or found in a book or magazine. To use a negative-acting photoresist with positive artwork, you have to use one of the techniques described in chapter 6 for converting a positive to a negative image. Positive-acting photoresists have the advantage that you don't have to convert the image in most cases.

Solvent-based versus aqueous processes

Both positive-acting and negative-acting photoresists can be solvent-based or aqueous. The two types differ in the type of developer required.

A solvent-based photoresist must be developed and stripped in an organic solvent. Xylene, trichloroethylene, and acetone are examples. Organic solvents are flammable and harmful when inhaled, so good ventilation is essential during their use. In addition, you must be careful to follow recommended procedures for their safe disposal.

Aqueous photoresists dissolve in a solution of water and sodium carbonate (washing soda or soda ash) or a similar water-based solution. Sodium carbonate solution is a mild, biodegradable caustic that is less toxic than the organic solvents, and health and safety precautions and disposal procedures for small quantities are less critical.

The use of less-harsh developers makes aqueous-based processes attractive. In addition, their thin, even coating of sensitizer allows you to make high-quality boards with narrow traces and close spacing.

A third category of photoresist, called *semiaqueous,* requires solvents for stripping, but not for developing.

Aqueous-process boards might cost a little more than solvent-process boards, but the overall cost for boards and chemicals often ends up about the same.

Presensitized boards or manual application

Another decision relating to photoresists is whether to buy boards that are precoated with sensitizer or whether to coat the boards yourself. Presensitized boards have a uniform layer of photoresist covering the copper. They are sold precut in convenient sizes and wrapped in lightproof bags, in both single- and double-sided versions.

Presensitized boards save you the trouble of applying your own photoresist, and they have a more even coating than you can apply yourself. Some aqueous-process resists are applied as a dry film and require special laminators.

Sensitizing your own boards does have some advantages. Sensitizing your own boards in large quantities is cheaper. With double-sided boards, you might want to develop and etch one side of a board at a time, and sensitizing your own boards allows you to do so. Overall, though, chances are good that you'll find the higher price of presensitized boards is well worth the convenience.

For do-it-yourself photosensitizing, spray-on and pour-on photoresists are available. If you decide to apply your own, the following sections describe what's involved in applying spray-on and pour-on photoresists.

Step-by-step: Applying spray-on sensitizer

This section describes how to apply GC Electronics' PCS aerosol positive-acting sensitizer.

Materials

You will need the following materials:

- PCS aerosol positive-acting sensitizer (GC Electronics).
- Copper-coated pc blank.
- Laboratory oven (optional). For drying and hardening the resist.
- Old newspapers or tarp to protect work surfaces while spraying.

Work area

Work in a well-ventilated room. Because the sensitizer is light-sensitive, room lighting must be subdued. A 40-watt incandescent bulb a few feet away from the pc board is acceptable. If it's daytime, be sure to cover all windows.

Relative humidity should be 35 percent or greater because the coating must absorb water molecules from the air. If the air is dry, use a humidifier or boil a pot of water and set it nearby.

Procedure

Follow these simple steps:

1. Clean the board, following the instructions in chapter 6.
2. Prop the board against a wall or other surface as close to vertical as possible. Lay newspapers underneath and in back of the board to catch the overspray (Fig. 7-18).
3. From a distance of 8 to 10 inches, spray the board in one continuous motion from bottom to top using a series of horizontal passes. Try to coat the board evenly.
4. When the board is coated, lay it face up in a drawer, cupboard, or other dark place where it can lie undisturbed to dry. Handle the board only by its edges and uncoated side. Use a paper towel to blot up excess coating that gathers at the edges.

    Dry the board overnight at room temperature, or you can speed up the process with heat. Let the board dry at room temperature for at least 15 minutes. Preheat a laboratory oven to 120°F to 125°F. Be sure the inside of the oven is dark. Remove any bulbs that turn on automatically when the door is opened. When the oven is preheated, turn the oven off and place the board in the oven for 20 to 25 minutes. After heating, allow the board to return to room temperature.

Fig. 7-18 *Setup for spraying sensitizer onto a copper-coated board.*

5. If you wish, you can now add a second coat of sensitizer, spraying and drying the board as before. Adding a second coat results in a more even coating overall. When the board is coated and dried, you're ready to expose and develop.

Flow-coating sensitizer

Flow-coating is another way of applying liquid sensitizer. In flow-coating, you pour the sensitizer onto a board, spread it across the board, and allow it to dry.

Kepro's KPR-3 is a negative-acting, solvent-based liquid photoresist that can be flow-coated by pouring a small amount on a board and tilting the board to cover it evenly. You can also use a brush to apply the coating. When dry, the sensitized board is ready for exposing.

Step-by-step:
Exposing artwork onto a board

When you have purchased or made your sensitized pc board, you're ready to expose the board with ultraviolet light. You must have a copy of your artwork on a transparency: a negative image for negative-acting photoresist or a positive image for positive-acting photoresist.

Before you expose an actual project, expose a small sample board as a test strip to find the best exposure time for your setup (Fig. 7-19). To make a test board, follow the instructions below, but expose different areas of the board for different times, as described under "Making a test strip" in chapter 6. It's also a good idea to make a test strip if you switch to a different type of sensitizer or change your light source.

A precisely measured exposure time isn't critical, but in general, you're better off overexposing than underexposing. With positive-acting sensitizer, underexposing means that a film of sensitizer remains on the background, preventing it from etching away. With negative-acting sensitizer, underexposing means that the pads and traces won't harden and will etch away.

Overexposing is a problem if the opaque areas on the transparency don't completely block light, because areas that should be protected from light will begin to expose slightly. Overexposing can also allow light to leak under the edges of the artwork, resulting in thinner traces and smaller pads on a positive-acting

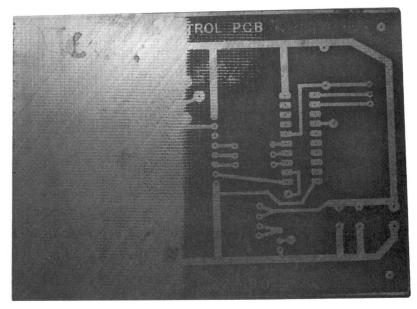

Fig. 7-19 *On this test board, different areas of the board received different exposure times to determine the best time to use with a particular setup.*

board. When you find an exposure time that gives good results, make a note of it for future reference.

Materials

You will need the following materials:

- The pc board with photosensitive coating. The following instructions describe how to expose Kepro's DF (dry film) negative-acting, aqueous-process boards, but other boards use similar procedures. Check the specific instructions that come with your boards or sensitizer for details.
- Artwork on a transparency. Use negative artwork for negative-acting boards and positive artwork for positive-acting boards.
- Ultraviolet lamp and mounting hardware. Possibilities include a 300-watt quartz halogen floodlight, #2 photoflood, 275-watt sunlamp, 15-watt fluorescent black light, or any light source that includes the range 350 to 425 nanometers. Even direct sunlight can be used. Stands are available for mounting the lamp over a flat surface, or you can rig your own setup using clamps or other hardware.

- Contact frame or a flat surface and a pane of glass. See chapter 6 for more about contact frames.
- Scissors.
- Clock or timer with a second hand.
- Resist pen or opaque ink. For touch-ups.

Work area

To avoid exposing the sensitized pc board, work under a yellow bug light or other subdued lighting.

Protect your eyes from the ultraviolet light. For maximum safety, use a remote timer to control the lamp.

Procedure

Follow these simple steps:

1. Under normal lighting, examine your artwork and fix any flaws. Fill in any breaks or pinholes with a resist pen or opaque ink (Fig. 7-20) and carefully brush or blow away any dust on the image. Also examine your contact frame or exposure glass and clean the surface if necessary.
2. Position the ultraviolet lamp so that the face of the bulb is about 12 inches above where the pc board will lie so that the lamp shines directly on the pc board. Don't turn the lamp on yet.
3. If you are using a contact frame, place it directly under the lamp.

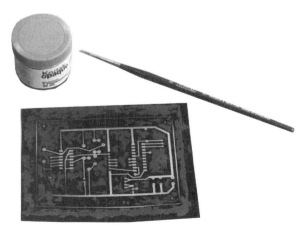

Fig. 7-20 *Use opaque ink to fill in pinholes or breaks in the artwork.*

4. Turn on your subdued lighting (if any), turn off the room lights, and cut open the lighttight bag containing the pc boards. Remove a board and reclose the bag to protect the remaining boards. The sensitized side of the board has a thin plastic film covering the sensitizer. Leave this film in place. Position the board, sensitized side up, in the contact frame or under the lamp.

5. Carefully position the transparency on the pc board. Be sure the artwork is oriented correctly. On a single-sided, through-hole board, you want the board to hold the bottom, or solder-side orientation, with DIP IC pads counting up clockwise around the chip. On a single-sided, surface-mount board, you want to expose the top, or component-side orientation, with SOIC pads counting up counterclockwise.

6. Close the contact frame or cover the transparency with a pane of glass. The glass presses the transparency firmly against the pc board and prevents light from reaching under the opaque areas in the artwork.

7. Turn on the ultraviolet lamp and expose the board for 12 to 14 minutes, or the exposure time you calculated from your test strip.

8. When the board has been exposed, wait at least 15 minutes before developing. If you need to turn on the room lights to prepare for developing, cover the exposed board or place it in a lighttight bag before you do so.

Step-by-step: Developing the image

Developing an exposed, negative-acting board dissolves the coating on the unexposed areas of the board and leaves the artwork as the etch resist. On a positive-acting board, developing dissolves the exposed areas, with the same result: the artwork remains as the etch resist.

The following is the recommended process for Kepro's DF negative-acting, aqueous-process boards. Other developing processes are similar. Be sure to read and follow the specific instructions included with any boards and developer you use.

Materials

You will need the following materials:

• Sensitized, exposed pc board.

- Developer. Kepro's DF boards develop in a solution of sodium carbonate. The DFP-101 professional kit includes a small packet of developer for each board. Other types of boards might require different developers. Be sure to use the developer recommended for use with your boards. The type of developer will vary, depending on whether the boards use positive-acting or negative-acting resist, and whether the resist is solvent-soluble or water-soluble. Your best bet is to buy the sensitizer or presensitized boards and developer from the same source.
- Developing tray. At least ½ inch deep and large enough to hold the pc board. Sodium carbonate developer can use an aluminum, glass, or plastic tray. If you use a different developer, follow the vendor's recommendations for the tray material. For example, metal trays are not suitable for use with GC developer for boards coated with PCS sensitizer.
- Ferric chloride etchant and glass or plastic tray. For checking development progress (optional).
- 1-pint measuring beaker or cup.
- Rubber gloves.
- Plastic spoon or other plastic or glass mixing tool.
- Sink or tray large enough to hold the developing tray.
- Scissors.
- Sponge. For wiping the board during developing.

Work area

Until the board is developed, work in subdued light as described under "Exposing the pc board." To mix the developer and rinse the board, you'll need a source of hot water. Wear old clothes and rubber gloves.

Procedure

Follow these simple steps:

1. Place the developing tray in a sink or larger tray and add hot water to surround the developing tray and keep it warm.
2. Measure 1 pint of hot water from the tap and pour it into the developing tray.
3. Cut open the bag and pour the premeasured developer into the water and mix to dissolve.

4. If you are using etchant to check development, pour the etchant into its tray.

5. Under subdued lighting, remove the exposed pc board from the exposure setup.

   Kepro's DF boards have a thin, protective film over the sensitized surface. This film must be removed before developing. Beginning at one corner, carefully peel the protective film from the board and discard. Do not touch the newly exposed surface of the board.

6. Place the pc board, image side up, in the developing tray. Gently rock the tray for about 1 minute.

7. Wipe the surface of the board with the sponge to help remove the unexposed sensitizer.

8. Remove the board from the developer and rinse it in warm water.

9. Feel the surface of the board with your hand. Any slick areas need more developing. Return the board to the developer and concentrate on wiping the slick areas with the sponge. Rinse and recheck.

10. When the board appears to be developed, check it by placing the rinsed board in the ferric chloride solution for about 20 seconds. Remove the board and examine. The areas to be etched should be a uniform pink. Shiny or glossy areas require more developing. To continue developing, rinse the board and return it to the developer for about 30 seconds. When the board is developed and rinsed, it's ready for etching, and you can turn the lights on.

Screen printing

Screen printing is yet another method for transferring artwork onto a pc blank. Screen printing is most useful if you need to make many copies of a board, because preparing the stencil is time-consuming, but one stencil can be used to make many identical boards.

Screen printing is a craft in itself. If you're already familiar with screen printing processes, from printing T-shirts, artistic works, or other applications, you might find it easy to adapt your techniques to pc board-making. The exact printing process varies, depending on the equipment and materials you use. Kepro sells a kit with materials and a printing frame for screen printing pc boards. Art supply shops also carry an assortment of screen printing materials.

In general, to make a stencil for screen printing, you prepare artwork on a transparency and expose it onto a sheet of film with a photosensitive coating, much like you expose other photosensitive films. The stencil is made by placing the developed film onto a sheet of silk, synthetic, or wire mesh stretched in a wood, aluminum, or other rigid frame. The soft film emulsion, which holds the image of the artwork, is pressed into the small holes in the wire mesh. When the emulsion dries and hardens, the backing film is peeled away and you are left with the artwork's image stenciled onto the wire screen.

To coat a pc blank with photoresist, fit the stencil tightly over the bare pc board, pour a liquid photoresist onto the stencil, and use a squeegee to force the resist ink into the holes in the stencil. The result is the pc board artwork drawn in resist ink on the pc board.

New methods

Two new image-transfer techniques require no etching. One uses photosensitive solder-mask film and conductive ink to transfer the image onto bare fiberglass base material. To use the method, you laminate solder mask onto the base material, cover with positive artwork, expose, develop, bake, and reexpose to cure the solder mask. The result is a negative image, with bare board where the pads and traces will be. To form the pads and traces, you use a squeegee to force conductive ink into the bare areas, and bake to harden. Think & Tinker is one source for more information about this process.

The second new method uses a computer-controlled engraver to draw the artwork by removing unwanted copper from a copper-coated board. The engraver works directly from the artwork files created by board-layout programs. Prices for these machines are currently around $12,000.

Image transfer for double-sided boards

When making double-sided boards, you must be sure that the top and bottom images register accurately, one over the other. In a perfectly registered board, a hole that passes through the center of a pad on the top of a board emerges exactly where it's supposed to: in the center of the corresponding pad on the bottom.

As a reference between the top and bottom layers, the artwork must contain matching alignment pads in at least three widely spaced locations in both the top and bottom artwork. On the board, a hole is carefully drilled in the center of each align-ment pad, and these holes are used to correctly position the art-work on the board.

There are several techniques you can use to register the top and bottom artwork, including registering before etching, regis-tering after etching one side, and registering after etching both sides. The following sections describe each of these methods.

Step-by-step: Aligning after etching

You can create a double-sided board by transferring the top and bottom artwork to individual, half-thickness, single-sided boards, etching the boards individually, and then registering and laminating the two boards together to form a single double-sided board. This is a simple and effective way to register the artwork.

Materials

You will need the following materials:

- Individual etched pc boards for the top and bottom layers. Boards $\frac{1}{32}$-inch thick work well. They are available in bare copper and presensitized boards, so you can use whatever image-transfer method you prefer. You can use General Consulting's copper-coated sheets for their Fast-Proto method. If you use the Fast-Proto sheets, one of the layers should be laminated to a piece of base material, as described earlier in this chapter.
- Alignment pins (Fig. 7-21). The alignment pins are thin, vertical pins on flat bases. The pins are used to align, or register, the layers on double-sided boards. Use three or four for small to medium boards; larger boards might re-quire more.
- Adhesive. General Consulting uses 3M's #9469 high-strength, silicon-base contact adhesive for laminating the copper to the base material. The adhesive comes on 3-inch wide rolls with a backing paper and can withstand solder-ing temperatures and solvents.
- Handheld rubber roller. Available at photo supply shops.
- Drill press and drill bits. See chapter 9 for more on drilling techniques and equipment.

Fig. 7-21   *Alignment pins for registering the top and bottom artwork on a double-sided board.*

Procedure

Follow these simple steps:

1. Cover one piece of the base material with the pressure-sensitive adhesive tape and use a roller or similar tool to press the adhesive onto the base material.
2. Drill the alignment holes—at least three widely spaced holes that appear in both the top and bottom artwork. Take your time in drilling. Before drilling, use a center punch or another sharp tool to indent the center of the alignment holes and help guide the drill bit. Drill as accurately as you can. Match the hole size to the diameter of the alignment pins so the pins will fit without slop. If the holes are accurately placed, the layers will register perfectly. Drill matching holes in both the top and bottom artwork.
3. On the piece of base material containing the adhesive, remove the backing tape and slide the base material, artwork side down, onto the alignment pins.
4. Carefully slide the remaining etched sheet onto the alignment pins and press it evenly onto the base material (Fig. 7-22).
5. Remove the alignment pins and use the roller to press the two layers together.

Fig. 7-22 *Aligning the layers of a double-sided board.*

Step-by-step:
Aligning before etching

An alternative to the previous method is to align the top and bottom artwork when you transfer the images before etching.

Materials

You will need the following materials:

- Materials for any image transfer method, but presensitized boards are not recommended because drilling these without damaging the coating of sensitizer is difficult.
- Drill press and drill bits (see chapter 9).

Procedure

Follow these simple steps:

1. Mark and drill alignment holes on the bare copper-coated board.

2. Using your method of choice, transfer the images to the board, one at a time, positioning the alignment pads over the drilled holes as you place the artwork on the board.

Step-by-step:
Aligning after etching one side

Instead of transferring both images before etching, an alternate method is to transfer and etch one image at a time. With this method, you can inspect one etched side before going on to the next. Because you transfer the first image before you drill the alignment holes, its placement isn't critical. You only have to worry about aligning the second image correctly.

Materials

You will need the following materials:

- Materials for any image transfer method, but do not use presensitized boards.
- Drill and drill press (see chapter 9).
- Acrylic or other etch-resistant paint or adhesive tape.

Procedure

Follow these simple steps:

1. Protect one side of the bare board by painting it or covering it with adhesive tape.
2. Use your method of choice to transfer the image to the uncovered side (iron-on, contact print, etc.).
3. Etch the board. Only the uncovered side will etch.
4. Drill alignment holes.
5. Paint or tape over the etched side of the board.
6. Remove the adhesive tape or paint from the previously covered side and clean the surface thoroughly.
7. Transfer the second image using the alignment holes as a guide.
8. Etch the board. This time the other side of the board will etch.
9. Remove the paint or tape from the first side.

REQUEST FOR QUOTE

DATE:_____

NAME OF COMPANY: _____
ADDRESS: _____
_____
CONTACT: _____
TELEPHONE: _____ FAX: _____

BOARD NAME/PART #: _____ REV: _____
QUANTITY: _____
DELIVERY REQUIRED: _____

TYPE OF BOARD: _____ SINGLE-SIDED _____ DOUBLE-SIDED
_____ MULTILAYER # OF LAYERS_____
BOARD DIMENSIONS: _____ X _____
MATERIAL THICKNESS: .031 .045 .062 .092 .125 _____
MATERIAL TYPE: FR4 TEFLON _____
COPPER OUNCES: 1/0 2/0 1/1 2/2 3/3 _____

TOLERANCES:
   MIN. LINE WIDTH: _____
   MIN. SPACING: _____
VIAS: (FOR MULTILAYER)
   BLIND: Y OR N _____
   BURIED: Y OR N _____
HOLES:
   QTY: _____
   # OF SIZES: _____
   SMALLEST HOLE: _____

SOLDERMASK: Y OR N _____
   TYPE: WET SCREEN PHOTOIMAGEABLE DRY
   SMOBC: Y OR N _____
   BOTH SIDES: Y OR N _____
SILKSCREEN: Y OR N _____
   COLOR: WHITE YELLOW BLACK _____
   COMP. SIDE ONLY: Y OR N _____
GOLD REQUIRED: Y OR N _____

ELECTRICAL TESTING: Y OR N _____
SMT: Y OR N _____

WHAT WILL YOU BE ABLE TO PROVIDE US WITH?
1:1 ARTWORK 2:1 ARTWORK GERBER FILE DRILL DISC DRILL TAPE BLUEPRINT

Fig. 7-23 *Quote sheet for commercial fabrication of pc boards.*

Contracting out for fabrication

As a final option, instead of fabricating your own boards, you can let someone else do the image transfer, etching, and drilling for you. Commercial fabricators can produce boards with fine traces, close spacing, and multiple layers. Disadvantages include the delay between requesting and receiving your board, and the generally higher cost per board for small orders. Fig. 7-23 is an example of a quote sheet that you might be asked to fill out for an estimate of costs for commercial fabrication.

Many fabricators can work from a variety of artwork sources, including actual-size or larger printed artwork and photoplotter and drill files on disk. Most will be happy to answer any questions you have and provide you with an estimate. Many of the magazines listed in appendix C are good sources for advertisements for board fabricators.

8
Etching

In the etching step, the final pc board is created by removing exposed copper, leaving the artwork drawn in copper on an otherwise bare board. Before etching, you need a copper-coated board with the artwork drawn in an etch-resistant medium such as an etch-resistant ink, developed photoresist, or transfer tapes and patterns. The artwork can be transferred to the board using any of the methods described in chapter 7.

To etch, immerse the pc board in an *etchant,* a chemical bath that removes the exposed copper from the board. The etched board is then ready for cleaning, drilling, component inserting, soldering, testing, and use. This chapter describes the etching process, including choosing an etchant, etching materials, and etching techniques.

Etching basics

The etchant must remove exposed copper on the pc board's surface, while leaving the etch resist and the copper beneath it untouched. During the etching process, a chemical reaction causes the copper to dissolve in the etching solution.

For etching in small batches, two popular etchants are ferric chloride and ammonium persulfate. Other persulfate etchants include sodium and potassium persulfate. Occasionally an image transfer method requires a particular etchant, but for most processes, the choice is yours.

Ferric chloride is usually sold as an orange liquid, although crystals that you dissolve in water are also available. The solution turns a dark, opaque brown as it absorbs copper during etching.

Ammonium persulfate is sold as crystals for dissolving in water. The solution starts out clear, but turns an increasingly deeper blue as it absorbs copper.

Fig. 8-1   *Two etchants: ammonium persulfate in crystal form (left) and ferric chloride solution (right).*

Figure 8-1 shows etchants as purchased in crystal and liquid form. Products marketed as etching chemicals might contain small amounts of additives for better performance, such as wetting and antifoaming agents.

In etching, faster is better. Slow etching allows the etchant to undercut, or remove copper under the edges of the etch resist, resulting in a lower-quality etched image. Three ways to speed up etching include mild heating, aerating, and agitating the etchant. Heating and aerating encourage oxidation, which is part of the etching process. Aerating and agitation ensure that fresh etchant continually washes over the exposed copper.

Check with local officials on how to dispose of used etchant. Etchants are caustic to metal plumbing, so when cleaning up and rinsing etching trays, be sure to flush the drains with generous amounts of water.

Step-by-step: Ferric chloride etching

Your etching setup can be as simple as a glass or plastic tray agitated by hand, as the following instructions for etching with ferric chloride show.

Materials

You will need the following materials (Fig. 8-2):

- A pc board with etch-resistant artwork image.
- Glass or plastic tray, at least 1 inch deep and large enough to hold the board to be etched.
- Sink or tray larger than the etching tray. For warming the etchant.
- Rubber gloves.
- Plastic or bamboo tongs.
- Ferric chloride etchant. One pint of ferric chloride etchant will etch 1 to 2 square feet of pc board coated with 1-ounce copper.
- Glass or plastic funnel.
- Abrasive pads or recommended solvent. For removing etch resist.

Work area

Work in a well-ventilated room. Wear old clothes and rubber gloves. Avoid letting the etchant touch your skin or eyes. Protect work surfaces from spills because etchants are caustic and will stain.

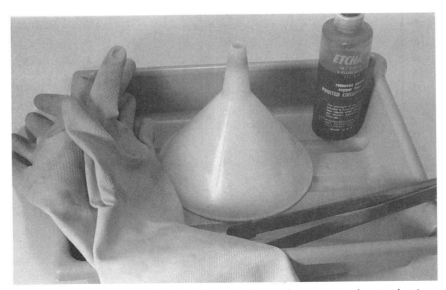

Fig. 8-2   *Etching materials include a plastic or glass tray, etchant, plastic or bamboo tongs, a funnel, and protective gloves.*

Procedure

Follow these simple steps:

1. Inspect your artwork on the board carefully. Use a resist pen or permanent marker to fill in any pinholes or breaks in the artwork. Use an artist knife to scrape off any etch resist that doesn't belong.
2. If your etchant is in solid form, follow the recommended procedure for dissolving it in water. Use only plastic or glass bottles and mixing tools.
3. Preheat the etching solution. A simple and safe way to heat the etchant is to fill a pail with hot water from the tap, place the bottle of etchant in the water, and wait about 10 minutes for the etchant to warm to the water's temperature. To prevent harmful fumes, do not heat the etchant beyond 110°F to 115°F.
4. After preheating, pour the etchant into your tray. To keep the etchant warm, place the etching tray in a larger tub or tray and fill the larger tray with hot water.
5. With the tongs, carefully place the pc board in the etchant (Fig. 8-3). For single-sided boards, you can etch copper side up or copper side down. Copper side down etching is faster because the used etchant falls away from the board allowing fresh etchant to contact the copper.

Fig. 8-3 *Placing the pc board in the etchant.*

Fig. 8-4   *Agitate the etchant by gently rocking the tray.*

If you etch copper side down, you must protect the etch resist from scratching, tearing, or falling off as it rubs against the bottom of the tray. Supporting the edges of the board with nylon (not metal) washers or standoffs will keep the board from contacting the tray bottom. For even etching of double-sided boards, give each side equal time face down.

6. Agitate by slowly rocking the tray (Fig. 8-4) or by using the tongs to repeatedly lift the pc board and return it to the etchant. The goal is to keep fresh etchant moving across the exposed copper, and to continually introduce air into the etchant. Be careful not to splatter or spill as you agitate.

   Eventually, you'll see the bare board begin to show through the copper (Fig. 8-5). Once the bare board begins to show, the etching should finish within 15 minutes. If a portion of the board seems to be etching slowly, concentrate on agitating the etchant over this section.

   Tray etching should take 10 to 30 minutes. If it takes longer than this, the etchant is too cold, it needs more vigorous agitation, or it can hold no more copper and needs replacing.

7. When all areas of bare copper have been etched, remove the board from the etchant and rinse it thoroughly in water.

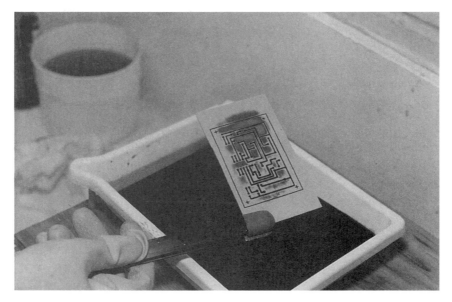

Fig. 8-5   *A partly etched board.*

8. Ferric chloride can be stored for reuse. Use a plastic or glass funnel to pour the etchant back into its bottle.

Etching time increases dramatically when the etching solution contains 8 to 11 ounces of copper per gallon. If etching begins to take much longer than it used to, it's time for fresh etchant.

9. Rinse the tray, tongs, funnel, and anything else that came in contact with the etchant using generous amounts of water.

10. To remove the etch resist that covers the artwork on the rinsed board, clean with a solvent that is recommended for use with your resist, or rub the etch resist off with abrasive pads. Remember to have good ventilation and wear protective gloves when using solvents.

Ammonium persulfate etching

Etching with ammonium persulfate is similar to etching with ferric chloride, with these differences

Fresh ammonium persulfate solution is clear, but turns a transparent blue as etching proceeds. Compared to the opaque brown of ferric chloride, it's easy to monitor boards that are etching in ammonium persulfate (Fig. 8-6).

Mix your etching solution in small batches, making just

Fig. 8-6   *Ammonium persulfate etchant starts out clear, then turns a transparent blue, which makes it easy to monitor the etching's progress.*

enough for immediate use. Ammonium persulfate is sold as crystals. Store the crystals in a tightly closed, airtight bag or container, and avoid high humidity. In solution, the etchant is unstable and should be used within 2 to 3 weeks. To ensure that the etching solution covers the pc board well, choose an etching tray just slightly larger than your pc board.

To mix, use a glass or plastic container and mixing tool, and dissolve about 4 ounces of crystals per pint of water. The etchant is usable until it holds 5 to 7 ounces of copper per gallon. One pint of solution will etch about 1 square foot of board.

Ammonium persulfate etchant works best when heated to about 115°F. As with ferric chloride, preheat the etchant by placing its container in a tub of hot water. To maintain the temperature while etching, place the etching tray in a larger tray of hot water. Don't overheat, because heating the solution to 150°F causes the etchant to decompose and become unusable for etching.

The etchant might also decompose if you place it in a poorly cleaned tray or container that you previously used for ferric chloride etching. If you use both etchants, use a separate tray and tongs for each.

Agitate, rinse, and clean your board as with ferric chloride. Save your etchant only if you plan to use it again in the next few weeks. Follow local recommendations for disposing of used etchant.

Problems and solutions

These are some of the common problems you might encounter in etching your boards, along with some recommended solutions:

- Etching takes too long. If etching takes longer than 20 to 30 minutes, your etchant is probably too cool or no longer usable. Be sure to preheat your etchant to the recommended temperature and keep it warm during etching. Use fresh etchant if you suspect your etchant is saturated due to previous use. Ammonium persulfate might also become unusable due to long storage or contamination.
- Etching is uneven. Once you begin to see the bare board during etching, the entire board should finish etching within about 15 minutes. To ensure even etching, concentrate on agitating the board evenly. Flip the board around to different orientations in the tray as you etch. For a double-sided board, give each side equal time face up and face down in the etchant.
- Some or all of the board does not etch. If you see no sign of etching after 30 minutes, something is preventing the etchant from reaching the copper. A common cause is photoresist that was incompletely exposed or developed. You can guard against this problem by testing your board after developing by briefly dipping it into the etchant, rinsing, and observing the results, as described in chapter 7. Exposed copper on a fully developed board will turn a dull pinkish-brown. A shiny blue or greenish coating means you need more developing time.
- Pads and traces have pinholes, breaks, or other open areas. This can occur if the transferred artwork is flawed to begin with, or if the etch resist comes loose from the board during etching. Before you etch, examine the etch-resist pattern on the board and use a resist pen or permanent marker to fill in any pinholes or breaks. If the etch resist lifts from the board during etching, the problem might be an imperfectly cleaned board or too little heat used with iron-on methods. A lightly adhering etch resist might be usable if you etch artwork side up.

Etching enhancements

There are several possibilities for making your etching more automated and efficient. For automatic agitation, you can design and build a motor-driven mount for the etching tray. The mount should slowly tilt one end of the tray up and down during etching, just enough to keep the etchant gently moving without spilling. An aerator with a plastic tube, such as the ones sold for aquarium use, will agitate the etchant and introduce oxygen.

If you do a lot of etching, you might want to invest in an etching tank that supports the boards vertically and heats and aerates the etchant. These are available from suppliers of pc board materials and equipment.

Step-by-step: Tin plating

After etching, boards can be tin plated to protect the bare copper and improve solderability. The etched, cleaned board is immersed in a plating solution that places a thin coating of tin on the exposed copper. The tin plating protects the copper from corrosion and gives it a shiny appearance. On boards with surface-mount components, the plating helps to form solid solder joints underneath the leads or terminations.

Copper that is freshly tin plated solders easily, but over time the tin can become very difficult to solder. It's important to solder a board soon after tin plating it.

An alternative to tin plating is to use a soldering iron to melt a thin coating of solder to coat the pads and traces in the artwork.

A popular product for tin plating of pc boards is Datak's TIN-NIT, which is sold as crystals. The following paragraphs describe how to use TINNIT to plate pc boards.

Materials

You will need the following materials:

- Etched, cleaned pc board.
- TINNIT crystals, from Datak Corporation. Enough to make 1 pint.
- Plastic or glass bottle, 1 pint.
- Tinning tray of heat-resistant glass or plastic.
- Rubber gloves.
- Glass or plastic beaker. For mixing.

- Mixing tool of glass or plastic.
- Funnel. For repouring into bottle.
- Bamboo or plastic tongs.
- Household ammonia. For board cleaning after tinning.

Work area

Work in a well-ventilated room. Wear old clothes and rubber gloves. Avoid contacting the skin or eyes with the tinning solution.

Procedure

Follow these simple steps:

1. Be sure the pc board to be tinned is clean. Residues will reduce the life of the tinning solution. Remove all etch resist with an appropriate solvent or abrasive cleaner and rinse thoroughly.
2. To mix the tinning solution, fill the mixing beaker with 12 ounces (1.5 cups) of hot water of about 130°F. Wearing protective gloves, dissolve the contents of the TINNIT package in the water. Add additional hot water to make 1 pint.
3. To keep the tinning solution warm, place the tinning tray in a larger tray filled with hot water, as recommended for etching.
4. Pour the tinning solution in the tray.
5. Immerse the pc board in the tinning solution (Fig. 8-7). Agitate occasionally.
6. After 10 to 30 minutes, use the tongs to remove the board from the tinning solution. Clean the board with a mild solution of household ammonia and rinse well.
7. Clean up. Pour the tinning solution into its bottle for storage. Shelf life of the solution is about 6 months.

Edge connectors

An *edge connector* is a series of conductive fingers along the edge of a pc board. The edge connector plugs into a mating series of contacts on a motherboard, backplane, or other pc board. Edge connectors must be wear resistant and highly conductive to withstand repeated inserting and removing and to ensure a good connection at each finger.

Edge connectors are etched onto the pc board as usual, then plated with nickel for wear resistance, followed by a layer of

Fig. 8-7   *Placing a board in a tinning solution.*

gold for good conductivity. The nickel and gold are added by electroplating, where electrodes are immersed in a bath and a voltage is applied between them. Current flows in the bath causing metal to be deposited on the more negative electrode, or cathode, in a process called *electrolysis*.

In electroplating, the edge connector acts as the cathode. The pc board pattern for an edge connector should include a plating bar, which is a trace, or bus, to which all of the fingers connect. This allows all of the fingers to be plated at the same time. After etching and plating, the plating bar is cut off, leaving the individual plated fingers.

Electroplating requires special equipment and materials, which are available from Kepro, PACE, and other suppliers.

Parts placement overlay

Hand-fabricated boards normally don't include a separate parts placement overlay on the pc board. The printed parts placement diagram serves as a reference for circuit construction and testing. In commercially fabricated boards, a parts placement overlay might be screen printed onto the board after etching. For hand-fabrication, you can use screen printing techniques or you can experiment with using the iron-on films described in chapter 7.

Solder masks

Many commercially made boards are covered by a *solder mask,* a coating that insulates and protects areas of the board that will not be soldered. A solder mask can prevent solder bridges from forming between traces on a board. Other reasons for adding a solder mask include minimizing and controlling cross talk and capacitive coupling between adjacent traces, and protecting against scratches, spills, contamination, and other hazards to exposed traces.

The solder mask might be screen printed onto the board or applied as a dry film that is laminated onto the board. Some masks require exposing to ultraviolet light and developing or heat curing.

For prototype and other hand-assembled boards, a solder mask is optional and is often omitted. For hand-application to selected areas, solder mask products are available for application by dipping, brushing, or applying by squeeze bottle. To suit different flux and cleaning requirements, there are solder masks of peelable latex and synthetic resins that are water and solvent soluble. Another solder mask option is Circuit Works' Overcoat Pen, which draws a line of protective overcoat material 0.02 to 0.03 inch wide (Fig. 8-8).

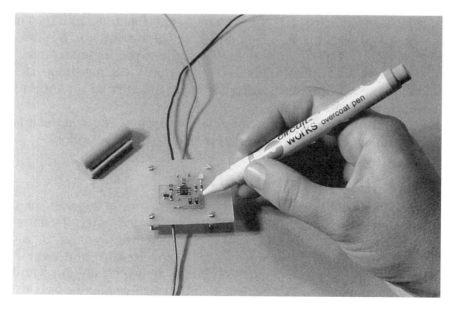

Fig. 8-8 *Using Circuit Works' overcoat pen.*

9
Drilling
and soldering

When you have an etched, cleaned pc board, you're ready to drill component and mounting holes, insert components, and solder. This chapter describes these steps.

Drilling component and mounting holes

Just about every board has some holes to be drilled, if only for mounting the board in its enclosure. Surface-mount designs often require holes for vias or a few through-hole components, such as connectors, switches, or other devices that are unavailable in surface-mount versions.

The goals in drilling a hole in a pc board are to place the drill bit accurately in the artwork and to drill through the board vertically, perpendicular to the board's surface. The hole should be clean, without entry or exit burrs. The right drilling equipment, drill bits, and techniques make a big difference in the quality of the holes you drill and how easy it is to drill them.

Drilling equipment

Drilling pc boards with a handheld drill is difficult because you must support the drill while keeping the bit vertical. If necessary, you can drill simple, noncritical boards with a hand drill.

A drill press will do a much better job, however, and with less effort. With a drill press, you need only place the board on the drilling plate, position the desired pad under the drill bit, and guide the bit down through the center of the pad. The drill press holds the bit vertically, and allows you to concentrate on positioning the board accurately as you guide the bit into it.

Two types of drill presses are widely used for hand drilling of pc boards. An ordinary benchtop drill press, such as might be used in woodworking or other general shop tasks, is adequate for occasional pc board drilling. An alternative to the benchtop press is a high-speed rotary tool with a drill press attachment. Figure 9-1 shows an example of this type of drill. The advantage of the rotary tool is its ability to drill at the high speeds recommended for pc board drilling.

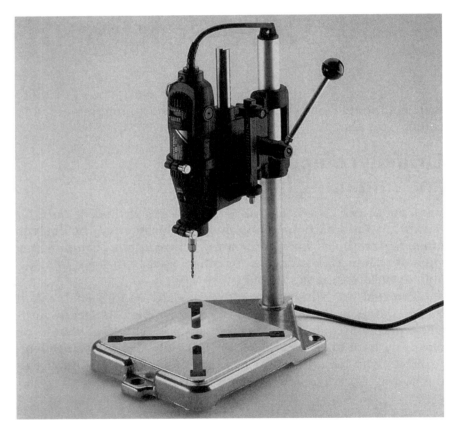

Fig. 9-1   *A high-speed rotary tool is useful for drilling through abrasive pc board base materials. (Dremel)*

Drill bits

A drill bit for pc board work must be hard enough to drill through abrasive pc board substrates without dulling, yet not so brittle that it breaks easily because of bit wobble or other stresses. High-speed steel drill bits are a good choice, and are available at moderate cost. Mail-order electronic surplus suppliers often carry resharpened bits at reduced prices. Low-cost carbon-steel bits will dull quickly when they are used to drill pc boards. Another option is to use high-priced tungsten-carbide bits. These bits are very hard, but also very brittle, and are prone to breaking if the bit wobbles or if the board shifts slightly during drilling.

Unlike the bits with fractional sizes of ⅛ inch, ¼ inch, and so on, the tiny bits used for pc board component holes are most often sold in numeric sizes, with higher numbers signifying smaller sizes. Table 9-1 shows bit sizes commonly used for pc board drilling.

Drill bits are available with straight or common shanks. The shank is the smooth portion of the bit that is gripped by the chuck. On a straight-shank bit, the diameter of the shank is the same as the diameter of the bit. On a common-shank bit, all bits have the same size shank.

When the shank is larger than the drilling diameter, the bit is an enlarged-shank bit. Common-shank bits for pc board drilling often have ⅛-inch diameter enlarged shanks. This size is easily gripped by the chuck on most benchtop drill presses. Figure 9-2 shows examples of pc board drill bits.

The choice of bit size depends on the diameter of the lead that will be inserted into the hole. A hole that fits the component lead exactly, with no slop, is easiest to solder, while a larger hole makes it easier to insert the lead, especially for ICs and other

Table 9-1   Numeric drill sizes.

Size	Diameter (inches)	Size	Diameter (inches)
80	0.0135	64	0.0360
79	0.0145	63	0.0370
78	0.0156	62	0.0380
77	0.0180	61	0.0390
76	0.0200	60	0.0400
75	0.0210	59	0.0410
74	0.0224	58	0.0420
73	0.0240	57	0.0430
72	0.0250	56	0.0465
71	0.0260	55	0.0520
70	0.0280	54	0.0550
69	0.0292	53	0.0595
68	0.0310	52	0.0635
67	0.0320	51	0.0670
66	0.0330	50	0.0700
65	0.0350		

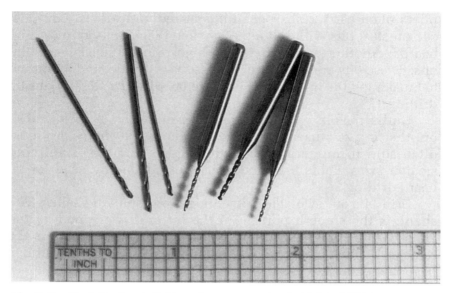

Fig. 9-2 *Straight-shank (left) and common-shank (right) drill bits.*

components with many leads that must be inserted at the same time. As mentioned in chapter 4, a hole's diameter should be between 0.006 inch and 0.020 inch wider than the lead inserted into it.

Remember that the minimum pad size increases with the hole size. For example, for a pad with an annular ring 0.02 inch wide, a 0.024-inch diameter hole requires a pad 0.064 inch in diameter, while a 0.050-inch hole requires a pad 0.090 inch in diameter.

For most projects, a few drill bit sizes will cover most requirements, as these examples show:

- #69 (0.0292-inch diameter). For vias that will hold AWG #30 (0.01-inch diameter) wire.
- #66 (0.033-inch diameter). For ¼-watt resistors and similar components.
- #61 (0.039-inch diameter). For DIP ICs.
- #55 (0.052-inch diameter). For TO-3 packages and other large-diameter leads.

A selection of fractional bits is also useful for drilling mounting holes. At a minimum, include ¹⁄₁₆ -, ⅛-, ¼-, and ⅜-inch drills in your toolbox. For mounting holes, precise bit size and hole placement are less critical, and you can drill and enlarge these holes by hand if necessary.

Drilling speed

The ideal drilling speed for a task depends on the bit's size and the material it is drilling. Harder materials call for faster drilling speeds. The optimum drilling speed for a material is specified as a surface speed in linear feet per minute, which describes the velocity of the outside edge of the drill bit. The recommended surface speed for FR-4 base material is 400 to 600 linear feet per minute.

Because smaller holes have smaller circumferences, to maintain the same surface speed for a given material, the drill bit must spin faster. To drill the tiny holes required by pc boards, drill speeds of 25,000 rpm (revolutions per minute) and higher are recommended.

On a benchtop drill press, the top speed available might be only 3000 rpm, but you can drill acceptable-quality holes even at this much slower speed. Rotary tools will drill at up to 30,000 rpm, and are a better choice if you do a lot of pc board drilling. A rotary tool is a handy piece of equipment with many uses beyond pc board drilling, including grinding, polishing, and carving, and these alternate uses might help you to justify the investment.

Step-by-step: Drilling a board

Following the procedure below will help you get good results when drilling pc boards.

Materials

You will need the following materials:

- Ready-to-drill pc board.
- Benchtop drill press or high-speed rotary tool with drill press attachment.
- Assortment of drill bits.
- Center punch or other sharp, pointed tool (optional). For marking holes.
- Safety goggles and dust mask.

Work area

Remember to take safety precautions when you drill. Wear safety goggles to protect your eyes and a dust mask to protect your lungs from fine particles of fiberglass and other materials. Do not wear loose clothing or jewelry. Read and heed your equipment's manuals.

Procedure

Follow these simple steps:

1. If your pads do not have etched center holes, use a center punch or a sharp pin or needle to indent the center of each pad to be drilled. The indentation will help to center your drill bit in the pad.

   A handheld center punch has a hardened, pointed tip for marking the center of a hole to be drilled. To use a hand punch, place its point in the desired location and tap the punch lightly with a hammer to make the indent. Spring-loaded punches don't require a hammer; just pressing down with the punch marks the hole (Fig. 9-3). Work carefully, because accurately placed indents will enable you to drill accurately placed holes with little effort.

2. When you are ready to drill, choose your drill bit and tighten it in your drill's chuck.

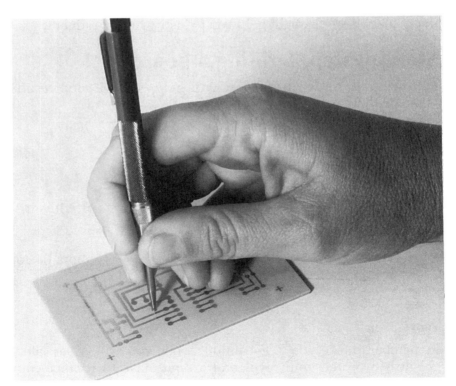

Fig. 9-3   *Using a center punch to mark a hole's center for drilling.*

3. Ensure that the drill is set to the desired speed. For most pc board component holes, this will be the highest speed available on your equipment.

4. Place the pc board on the drilling plate and, if possible, adjust the drill so that the tip of the bit rests ½ inch or so above the surface of the board.

5. Turn on the drill and visually inspect to be sure that the bit has no side-to-side wobble as it turns. If the bit is wobbling, stop the drill and inspect your setup. Reasons for wobble include a bent shaft on the bit, crooked installation of the bit in its chuck, or dust or particles in the chuck.

6. When the bit is spinning properly, lower it so that it almost touches the pc board and position the board so that the center of the pad to be drilled is directly under the bit. Carefully lower the spinning bit into the board to cut the desired hole. Be careful not to move the pc board while drilling. When the hole is drilled through, raise the bit out of the board.

7. When drilling your first few holes, remove the board and examine each hole after drilling. The holes should be cleanly made and centered in their pads. A dull bit might leave entry or exit burrs, which are rough edges on the top or bottom surface of the hole. Minor burring will cause no problems, but pads with smooth, even edges are easiest to solder. A piece of scrap base material or other rigid material placed under the pc board during drilling can help in eliminating exit burrs. If necessary, you can remove burrs with fine-grit sandpaper.

8. On a double-sided board, check to be sure that the holes emerge in the center of their corresponding pads on the bottom side. If they do not, the problem is either poor registration of the top and bottom artwork, or a drill bit that passed through the board at an angle, rather than vertically. A board with slight drilling imperfections might still be usable.

9. Drill all of your holes as described above, changing bits as needed. Be sure to use larger bits for any components with thicker leads. When you're finished, check the board carefully to be sure you didn't overlook any holes. Drilling a previously overlooked hole on a half-soldered board can be awkward or impossible, so be sure you haven't missed anything.

10. Don't forget to drill any holes needed for mounting the pc board in its enclosure. For these larger holes, you might find it easier to drill a small pilot hole first, and then enlarge it by redrilling with a larger bit.

Soldering tools and materials

When your board is drilled, you're ready to insert or place the components and solder them to the board. Hand-soldering involves applying heat and solder to the junction between a component lead or termination and its pad until the solder melts, forming the solder joint.

Proper choice of solder, soldering iron and other tools, and good soldering technique will ensure that the connections you solder are electrically, metallurgically, and mechanically sound. Through-hole and surface-mount components require different soldering techniques. The following sections describe how to choose and use soldering tools and materials for different types of projects.

About solder

Solder is the material that connects the component leads and terminations to their pads on the pc board. Solder is an alloy, or a mixture of metals, that, when melted, forms a metallurgical bond and an electrical connection to a base material. When solder is applied to a joint on a pc board and heated, it melts and connects the pad and lead. This connection is called the *solder joint.*

There are several reasons why solder is used to form electrical connections:

- Solder melts at a relatively low temperature (around 370°F), so heat damage to components is avoided.
- Solder bonds with the metals it connects. As the solder melts on a pc board, it flows evenly over the pad and its lead or termination and alloys with it to form a bond over a large surface area.
- Solder joints are reversible. To remove and replace a soldered component, you need only reheat the solder joints to melt the solder, lift the component leads from the board, and solder a new component in its place.

Electrical solder is usually an alloy of lead and tin. A popular general-purpose solder is 60/40 solder, which contains 60 percent tin and 40 percent lead and melts at around 370°F. Small amounts of other metals might be included as well, to improve performance.

The 60/40 solder has a small pasty range, from around 360°F to 370°F, where the solder is in a semisolid form. If a solder joint is disturbed while it's in its pasty range, the joint will be dull in appearance and weak in structure.

Besides 60/40 solder, other tin/lead ratios are available, each with a different melting point. One special type of solder is *eutectic* solder, which has a 63/37 ratio and the lowest melting point of all tin/lead solders—361°F. In addition, it is the only tin/lead solder that does not pass through a pasty range as it cools, but changes immediately from a liquid to a solid. Solders with larger or smaller ratios of tin to lead have higher melting points and larger pasty ranges.

Other tin/lead ratios include low-cost 50/50 solder and higher-cost 70/30 solder, sometimes used for pretinning. Special low-temperature solders are also available for soldering very heat-sensitive components.

Solder comes in many forms, including bars, preformed shapes, paste or cream, and wire. Most hand-soldering is done with wire solder, which is unwound from a spool as it is used. Solder in paste form is often used to solder surface-mount components.

Different diameter solders are available for different tasks (Fig. 9-4). Table 9-2 gives the diameters of different gauge sol-

Fig. 9-4 *Stock a selection of solders in your workshop for use on different types of tasks.*

Table 9-2   Diameter of wire solders.

Solder gauge	Diameter (inches)
10	0.128
14	0.080
16	0.064
18	0.048
19	0.040
20	0.036
21	0.032
22	0.028
24	0.022
28	0.014
32	0.011
34	0.009

ders. A good all-purpose size for soldering through-hole components is #22, which has a diameter of 0.028 inch. For soldering tiny SMT components, #28 solder, with a diameter of 0.014 inch, is a good choice.

Soldering flux

Inside most wire solder is a core or cores of flux, which help in forming the solder joint. Flux has several purposes:

- It promotes *wetting*—the ability of the solder to flow evenly across the surfaces to be joined.
- It loosens and dissolves metal oxides and other contaminants on the surface to be soldered, leaving a clean surface.
- It helps to transfer heat to the surface being soldered.
- It coats the joint to help keep it from oxidizing after soldering.

Soldering fluxes differ in how effective they are in dissolving contaminants and promoting wetting, and they also differ in the type of residues they leave after soldering. The residues determine the type of cleaning (if any) the board requires after soldering. Three major categories of fluxes are solvent soluble, water soluble, and no clean.

No-clean fluxes   No-clean fluxes leave no harmful residues and do not require cleaning after soldering. These are the most convenient, and should be your first choice for hand-soldering.

Multicore Solders offers several wire solders with no-clean fluxes. Xersin 2000 was their original no-clean solder, which leaves a visible but clear and harmless residue. Xersin 2005 is a no-halide version of this solder. Other advantages of these solders are much reduced fuming and smoking as compared to solders with rosin-based fluxes. X-32 and X-42 are no-clean solders that leave minimal flux residues, with X-32 being halide-free. Because of the very fast activation and decomposition rates of these solders, you must tin your tip with a special tip tinner/cleaner (TTCL). Traditional tinning will result in the solder balling up on the tip. X-38 is a further improved no-clean solder with a no-residue flux that does not require a special tip tinner.

Solvent-soluble fluxes   Solvent-soluble, resin-based fluxes have been popular for many years. Rosin is a naturally occurring resin that is extracted from pine trees and has long been used as a soldering flux. Synthetic rosin substitutes also exist. Most fluxes also include small quantities of activators, which are special substances that improve the flux's ability to dissolve metal oxides and promote wetting.

Three types of rosin flux are R (rosin), RMA (rosin mildly activated), and RA (rosin activated). Type R is the least active, which means it has the weakest cleaning ability, and its use is limited to easily soldered joints. Type RMA is more active, and RA is the most active of the three. Type RA flux is suitable for general-purpose soldering, while types RMA and R might be specified for use in military and other special applications.

Two other solvent-soluble fluxes are SRA (super rosin activated) and SA (synthetic activated). Both of these are mildly corrosive, and should be cleaned from the pc board after soldering.

Many rosin-based fluxes, including Multicore Solders' Ersin fluxes, are noncorrosive, which means that after soldering, the flux leaves no acid residues that will corrode or chemically wear away parts of the pc board or solder joint. Rosin-based fluxes are also nonconductive, which means that the residues will not introduce unwanted electrical connections on the board.

The pc boards that are hand-soldered with noncorrosive, nonconductive rosin-based flux often do not require cleaning to remove flux residues. However, some flux residues remain sticky or tacky after soldering, which might attract contaminants to the board if the residues aren't removed.

Most of the residues left after soldering with a rosin-based flux are nonpolar, or nonionic, types of contaminants. Examples of other nonpolar contaminants are oil, grease, and fingerprints.

In the past, cleaning nonpolar contaminants from a board soldered with a rosin-based flux required using solvents containing chlorinated fluorocarbons (CFCs), whose use is now restricted due to environmental concerns about damage to the atmospheric ozone layer. In response to the restrictions, new CFC-free cleaning formulations are appearing. Many of these substitutes use hydrochlorofluorocarbons (HCFCs) in place of CFCs. HCFCs have just 2 to 10 percent of the ozone-depleting effect of CFCs, and are approved for use by the EPA (Environmental Protection Agency).

Water-soluble fluxes   A third flux alternative is water-soluble flux, whose residues contain polar ions that can be removed with a polar, or ionic, solvent such as water. Water cleaning eliminates the need to use harsh solvents. But because the residues from water-soluble fluxes can become corrosive or conductive, thorough cleaning immediately after soldering is essential.

Multicore Solders' Hydro-X is an example of a wire solder with a core of water-soluble flux. Hydro-X is extremely active and can be used on even difficult-to-solder parts. The flux residues must be removed immediately after soldering by thorough washing in water or another aqueous cleaner.

In addition to water, aqueous cleaners include surfactants, such as soaps, which lower the surface tension of water; saponifiers, which are chemicals that change rosin into soap and also lower the surface tension of water; softeners, which increase the effectiveness of the surfactants; additives, which dissolve contaminants; neutralizers, which stop the action of the acid in the flux; and detergents.

Solder paste

Solder is also available in paste or cream form. Solder paste consists of tiny particles of solder suspended in a viscous flux. The same types of fluxes available in wire solder are also available in solder paste.

Solder paste is widely used to solder surface-mount components. The paste is applied to component pads, the component is placed on the pads, and heat melts the solder to form the solder joint. Solder paste can be applied by screen or stencil printing, or with a syringe or other hand applicator.

Soldering irons

The soldering iron is the tool that heats the joint and melts the solder. Choosing the right iron and soldering tip will make a job go more smoothly.

For hand-soldering of pc boards, you'll need a pencil-type iron, which is held much like a pencil. Figure 9-5 shows a pencil-type iron in a soldering station, which includes a power source, sponge tray, and sometimes a temperature control. Larger soldering jobs might require a soldering gun, which is shaped and held like a pistol. Soldering guns have larger tips than typical pencil-type irons.

The soldering iron and tip must be able to heat the solder joint and quickly melt the solder. In addition, after soldering a joint, the iron should recover quickly, with no noticeable wait for the iron to return to soldering temperature before soldering the next joint. How well the soldering iron performs depends on the

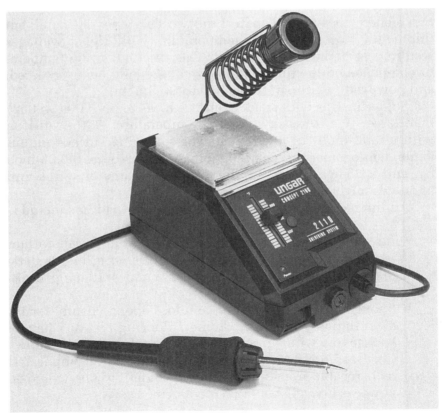

Fig. 9-5 *A typical soldering station. (Ungar)*

iron's capacity, its efficiency, the tip temperature, the tip size, and the size of the joint to be soldered.

Tip temperatures for soldering range from 500°F to 1000°F. A temperature of at least 370°F is needed to melt the solder. To this, add 150°F to 200°F for the heat the iron must transfer to the joint, plus a reserve for fast recovery, and the result is a typical tip temperature of 600°F to 800°F.

If the tip temperature is too high, components might be damaged and the foil might lift from the pc board. On the other hand, if the temperature is too low, soldering will take longer, and components might be damaged by prolonged exposure to the heat. For most soldering jobs, a temperature range for effective and safe soldering is easily found by experimenting.

On a temperature-controlled soldering iron, you need only select the desired tip temperature and the iron's internal control circuits maintain the temperature automatically. Other irons use interchangeable keys for soldering at different temperatures.

Many irons are rated only by wattage, which describes how much energy is available to heat the tip. This gives a general, but imprecise, indication of tip temperature, with higher wattages resulting in higher temperatures. The exact soldering temperature depends on the tip size, the size of the joint being soldered, and how well the iron transfers its heat to the tip.

Soldering tips with smaller surface areas require less wattage than larger tips to reach the same temperature. A 15-watt iron with a 0.02-inch tip is adequate for soldering surface-mount joints, while larger through-hole joints might require a 0.05-inch tip and a 25-watt iron. Soldering large-diameter wires and terminals might require a 40-watt iron.

Additional features of many modern soldering irons include

- Low-voltage operation (12 to 24 volts) and transformer isolation. These prevent component damage due to static electricity and voltage leakage when soldering high-impedance circuits.
- Sponge tray. Keep a damp cellulose sponge handy for tip cleaning as you solder. Some soldering stations have a sponge tray with a reservoir that keeps the sponge damp.
- Cordless operation. Butane-fueled or battery-operated rechargeable irons are handy for soldering jobs where ac power is unavailable or inconvenient.

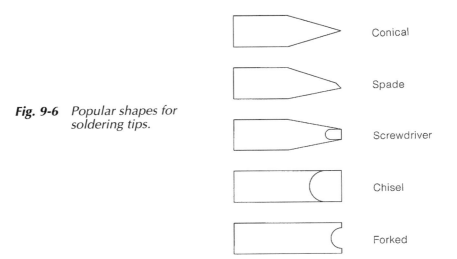

Fig. 9-6 *Popular shapes for soldering tips.*

Conical

Spade

Screwdriver

Chisel

Forked

Choosing soldering tips

Most soldering irons have removable tips, which allows you to choose a tip to suit the job at hand. Soldering tips come in many sizes and shapes: screwdriver, conical, chisel, and spade are popular ones (Fig. 9-6). Long, narrow tips are good for reaching into tight spaces, while angled tips can reach into corners. Some special tips for surface-mount devices are forked, with two or more tiny blades that heat and solder all of the connections on a device at once. Soldering tip sizes range from 0.01 inch on up.

Soldering tips are typically made of iron- or nickel-plated copper. The copper has good heat conduction, while a thick plating of iron or nickel prevents corrosion.

The ideal tip contacts the solder joint as completely as possible. A few carefully selected tips will handle most jobs. A basic, all-purpose set of soldering tips might include the following:

- 0.03-inch or $\frac{1}{32}$-inch screwdriver. For through-hole soldering of DIPs and other components.
- 0.02-inch or $\frac{1}{64}$-inch screwdriver or spade. For surface-mount and other fine soldering.
- 0.07-inch or $\frac{1}{16}$-inch screwdriver or chisel. For medium-duty soldering to wires and terminals.
- 0.13-inch or $\frac{1}{8}$-inch chisel. For heavy-duty jobs.

You can adapt the above list to suit your own needs, adding other special-purpose tips as you need them.

Soldering components to a board

When you solder components to a board, begin with the least sensitive or fragile components and work up to the most sensitive ones. In general, use this order of insertion or placement and soldering:

- IC sockets (the sockets only, not the ICs) and connectors.
- Resistors, capacitors, and other passive components.
- Unsocketed ICs, crystals, and other active or delicate components.
- Wires, cables, or other bulky items that are impractical to solder earlier.

Step-by-step: Soldering a through-hole board

Through-hole components are easily soldered by hand. The following section describes how to solder a DIP IC, resistor, and other components to a board.

Materials

You will need the following materials:

- The pc board and components to be soldered.
- Soldering iron with 0.03-inch screwdriver or similar tip.
- Solder, 60/40 tin/lead ratio, #22 gauge, or similar.
- Needle-nose pliers. For lead bending.
- Wire cutters.
- A pc board holder or "extra-hands" tool (optional).

Work area

Solder only in a well-ventilated room. Use a vent hood or fan to carry solder fumes from your work area. Do not eat or drink while soldering. Immediately after soldering, wash your hands to avoid ingesting lead. Good lighting is also important. Use a task lamp to illuminate your work area.

Procedure

Be sure the pc board artwork is clean. If the board has fingerprints, dirt, or other contamination, or if much time has passed since the board was etched, it's a good idea to clean the artwork with fine steel wool or another mild abrasive before soldering.

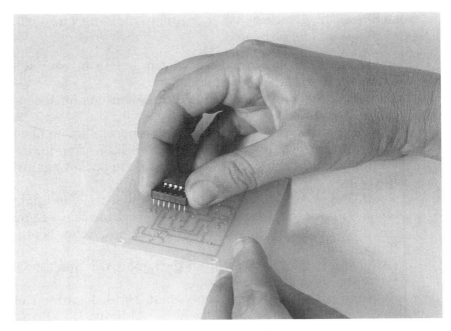

Fig. 9-7   Installing a DIP IC onto a pc board.

Fig. 9-8   Board-holding tools hold the pc board steady during soldering.

Soldering a DIP IC socket We'll begin by soldering a through-hole DIP IC socket to a pc board:

1. Install a 0.03-inch screwdriver or similar tip in your soldering iron and turn the iron on.
2. Insert the leads of the socket into their holes on the pc board (Fig. 9-7).
3. Turn the pc board over and use needle-nose pliers to bend two opposite legs (pins 1 and 14 of a 14-pin socket, for example) slightly outward to help hold the socket against the board.
4. Set the board in a stable position for soldering. You might be able to rest the board on a table or work surface, or you can grip it in a vise, an extra-hands tool, or another board-holding tool (Fig. 9-8).
5. Unwind a strand of solder from the spool. You don't have to cut the strand from the spool.
6. Grip the soldering iron like a pencil. Hold the strand of solder in your other hand. Place the soldering iron tip so that it contacts both the pad and the lead to be soldered. Place the end of the strand of solder against the joint on the other side of the lead.
7. In ½ second or so, the pad and lead will heat up, and the solder will melt to form an electrical and metallurgical connection between the pad and lead.
8. When the solder has flowed over the lead and pad, remove the strand of solder, then the soldering iron, and allow the joint to cool undisturbed for at least a few seconds.

The joint should appear clean and shiny, showing a smooth flow of solder. Figures 9-9 and 9-10 show examples of good and bad solder joints.

A dull solder joint indicates that the solder was disturbed as it cooled. To repair the joint, reapply the soldering iron to remelt the solder, remove the iron, and allow the joint to cool undisturbed.

Solder that balls around the joint, rather than flowing smoothly over it, is an indication that the soldering temperature or time is too low, or that the lead or pad is contaminated and the solder can't penetrate to the joint. Fine sandpaper or steel wool will clean a dirty pad or lead.

After soldering the first joint of an IC socket, check to be sure that the socket is resting flat against the board. If it isn't, reheat the joint while pressing the socket against the board from the other side. As the solder remelts, the

Fig. 9-9   *Too little solder and heat resulted in solder balling up and not covering the pads.*

Fig. 9-10   *A proper amount of solder and heat results in smooth, shiny solder joints that cover the pads.*

socket will be free to shift into position. Remove the iron and again wait a few seconds for the joint to cool.

9. After the first joint is soldered, solder the rest of the leads in the same way.

10. When you finish the first socket, solder the other sockets in the circuit. While soldering, clean your iron's tip periodically by wiping it against a damp sponge.

Soldering a resistor   To solder a through-hole resistor to the board, use needle-nose pliers to bend the leads 90 degrees so they will fit in their drilled holes (Fig. 9-11). Some resistors come with prebent leads. Insert the leads in their holes on the board. Bend the leads slightly underneath the board to help hold the resistor in place. If the resistor will dissipate heat in the circuit, leave a small amount of clearance between the resistor's body and the board. Otherwise, the resistor body can rest against the pc board.

Solder the joints in the same way as you soldered the DIP IC. Touch the soldering iron to the joint, apply solder to the opposite side of the joint, wait for the solder to melt and flow over the joint, and remove the solder and soldering iron (Fig. 9-12). After soldering, trim long leads with wire cutters (Fig. 9-13). Solder capacitors, transistors, and other components with bendable leads in the same way.

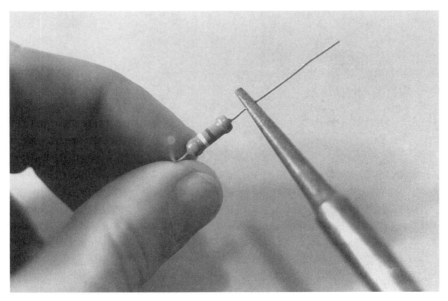

Fig. 9-11 *Bending a resistor's leads for pc board insertion.*

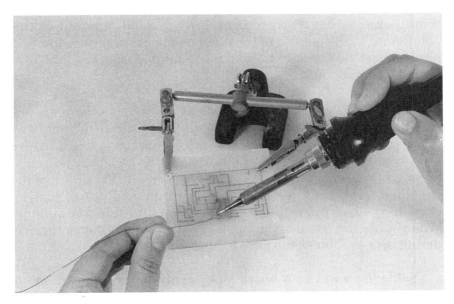

Fig. 9-12   *Soldering a through-hole resistor to a pc board.*

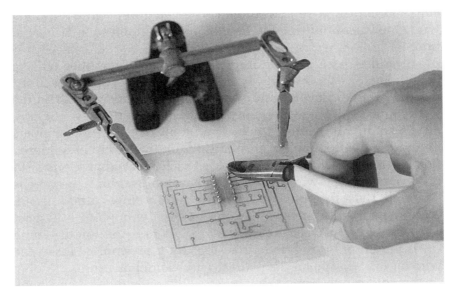

Fig. 9-13   *Trimming leads.*

Installing heat sinks   If a component will be mounted with a heat sink, heat sink compound ensures good heat conduction between the component and its heat sink. Tubes of heat sink compound are available from electronics parts suppliers. Use a toothpick or other applicator to apply a small amount of the compound to the bottom of the component, press the component onto its heat sink, and use a screw and nut to fasten the heat sink onto the component.

Soldering heat-sensitive components Some components are easily damaged when exposed to soldering temperatures. For these components, use a heat sink while soldering to divert the heat from the components. The heat sink can be any conductive mass you can clamp onto the component body. Possibilities include heat sink clips designed for this purpose, alligator clips, or even long-nose pliers held or clamped onto the component. When soldering sensitive components, be sure to remove the soldering iron as soon as the joint is formed.

Soldering wires to a board On many pc boards, off-board components connect to the board with individual, insulated hookup wires that are soldered to pads on the pc board. The wires can be solid or stranded. Solid hookup wire consists of a single strand of wire surrounded by insulation, while stranded wire consists of multiple bare wires in a common tube of insulation.

Stranded wire is more durable for connections that will be flexed, such as probe wires or external cabling. For connections that will remain largely undisturbed after soldering, such as the wiring inside an enclosure, you can use solid wire.

To solder an insulated wire to a pc board, use wire strippers to remove about ¼ inch of insulation from the end to be soldered (Fig. 9-14). The wire end will solder easier if it is pretinned. To do this, clamp the wire in a vise or extra-hands tool. If the wire is stranded, twist the strands together to form a single, thick strand. Melt a small amount of solder on the wire, enough to fully coat the exposed end (Fig. 9-15). Some wires are pretinned, which saves you from having to do this step. After pretinning, insert the wire end in its hole on the pc board and solder as usual.

To solder two wires together, strip the ends of both and hook the two ends around each other to form a mechanical connection. Use a vise or clamp to hold the wires in place, and apply a hot soldering iron and solder to make the solder joint.

To solder a wire to a terminal on a switch or other panel-mounted component, loop the stripped end through or around the terminal and apply a hot soldering iron and a liberal amount of solder to coat the joint.

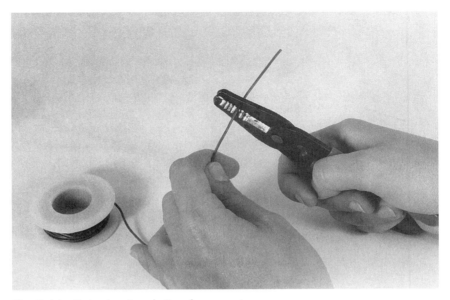

Fig. 9-14   *Stripping insulation from a wire.*

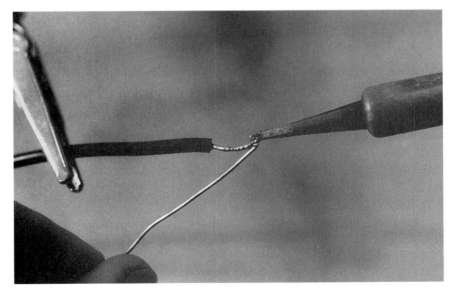

Fig. 9-15   *Tinning the end of a wire.*

You can use heat-shrink tubing to cover and insulate bare connections on wires or terminals. Insulation is especially important for high-current or high-voltage connections that could cause physical harm if someone accidentally touched them or shorted them to ground. Heat-shrink tubing is a flame-retardant

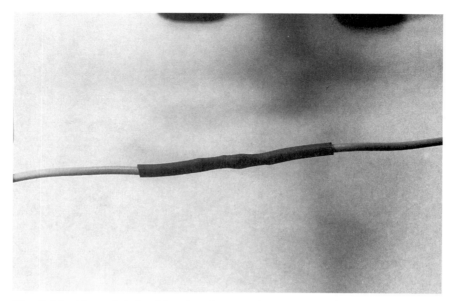

Fig. 9-16 *Heat-shrink tubing insulates and protects solder joints.*

polyolefin tubing that permanently shrinks 50 percent in diameter when heated. Figure 9-16 shows a connection covered by a length of heat-shrink tubing. The tubing is available in a variety of sizes. Choose a size that fits comfortably over the joint to be insulated. Cut a piece long enough to cover the joint with generous overlap, and slide the tubing over the joint. Be sure to slide the tubing on before you solder both ends of the wires.

When you heat the tubing with a hot-air source, butane lighter, or match, its diameter shrinks to the diameter of the wire and joint it surrounds. The result is a tight-fitting, insulating cover for the previously bare joint.

Finishing up   When you're done soldering for the day, wipe the tip of the soldering iron to clean it, apply a generous coating of solder to protect the tip, and turn off the iron.

Installing ICs in their sockets   DIP ICs that are fresh from the manufacturer usually require some minor lead bending before they will fit in the holes in their sockets or on a pc board. Place the IC on a flat surface so that one row of legs lies flat on the table's surface. Rock the IC slightly and press to bend the row of legs so that they are perpendicular to the top of the IC package. Do the same for the other side. Try to insert the socket and repeat as necessary until the IC fits into its holes.

Step-by-step:
Surface-mount soldering

Soldering surface-mount components requires some different techniques than through-hole soldering. The main differences result from the tiny size of surface-mount joints and their lack of through-holes to hold the components in place. Their small size also means that they can be more heat sensitive than their larger, through-hole equivalents. To hand-solder surface-mount joints, you can use wire solder or soldering paste. The following sections describe hand-soldering of surface-mount components using both of these materials.

Tools for surface-mount soldering

Special tools and materials for hand-soldering of surface-mount components include tweezers or a vacuum tool, a magnifier, and adhesives to hold the components in place.

Tweezers   Simple tweezers will do for occasional placement of surface-mount components. Electronic supply catalogs have a huge variety of specialized tweezers with fine, medium, and wide tips in straight, curved, and angled configurations. Tweezers are available in stainless steel, plastic, and ceramic materials.

To lift a component with a tweezers, grasp the component by its body, not by the leads or terminations, and squeeze gently, just hard enough so that you won't drop the component (Fig. 9-17).

Another option for component placing is vacuum tweezers, also called a vacuum pencil, vacuum pickup, or vacuum pipette. This is a narrow, hollow, rigid tube with a vacuum source at one end and a vacuum probe or cup on the other. When you press the probe onto a component's body, the vacuum holds the component against the probe, allowing you to lift and move the component. To release the component, you release the vacuum.

Some vacuum tweezers include an electrically powered vacuum generator. Others have a squeezable bulb that creates light suction without external power.

Figure 9-18 shows ESP Solder Plus' VAC Tweezer, which is a low-cost, hand-operated vacuum tweezers. The VAC Tweezer comes with a series of changeable tips and suction cups to fit different-size components. To lift a component with the VAC Tweezer, you install the appropriate tip, squeeze the bulb, press the tip onto the component, and release the bulb to create suction that holds the component against the tip (Fig. 9-19). You can

Fig. 9-17   *Lifting a surface-mount resistor with tweezers.*

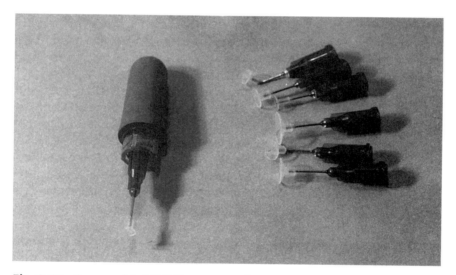

Fig. 9-18   *Squeezable VAC Tweezer and tips.*

now move the component to its desired location, where you squeeze the bulb again to release the component from the tip.

Magnifier   A tabletop magnifier will give you a better view of your work and eliminate eyestrain. If you don't have one of these, at least have a hand magnifier handy to inspect your work after soldering.

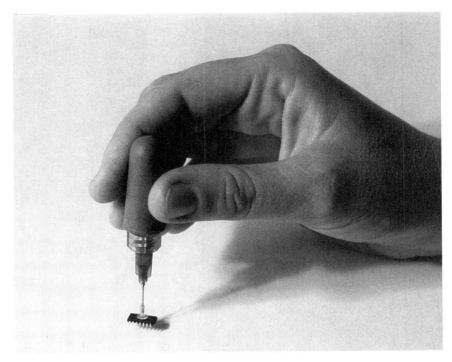

Fig. 9-19 *Lifting a surface-mount IC with a vacuum tweezers.*

Adhesives   Because surface-mount devices have no through-holes, you have to find another way to anchor the components to the board while soldering. After soldering, the strength of the solder joint holds the small, light components in place.

The material used for anchoring surface-mount components must be noncorrosive. Alternatives include adhesive applied underneath the component body, tape temporarily applied over the component body, and solder paste applied to the component pads. For hand-soldering, the adhesive needs to hold just well enough to enable you to solder the first joint on the component. Suitable adhesives for component bodies include epoxies, acrylics, and urethanes. Use a syringe, toothpick, or other pointed tool to place a tiny dot of adhesive on the pc board where the component's body will rest.

A simple method of component anchoring uses masking tape, or any light-holding tape that doesn't leave a residue on the board. Cut a strip of tape and press the edge of the tape onto the component's body, leaving at least one lead or termination exposed. With the component attached to the tape, position the component on its pads on the pc board and press the tape onto

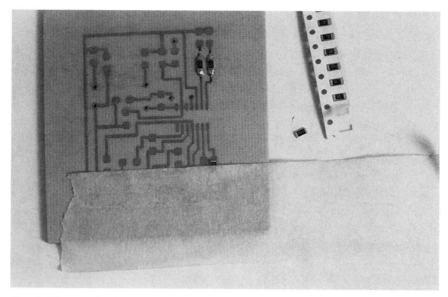

Fig. 9-20   *Anchoring a component with masking tape.*

the board to hold the component in place (Fig. 9-20). The tape can be removed after soldering at least one connection.

Another option is to apply a dot or thin layer of solder paste to the component pads, using the paste to hold the component in place during soldering, as described later in this chapter.

Step-by-step: Soldering a surface-mount board with wire solder

The following section describes how to solder surface-mount components to a board using a soldering iron and wire solder.

Materials

You will need the following materials:

- The pc board and surface-mount components to be soldered.
- Solder, 60/40 tin/lead ratio, #28 gauge (0.014-inch diameter) preferred.
- Soldering iron with 0.002-inch screwdriver tip or similar. Use a temperature-controlled or low-wattage (15 watts) iron.
- Hand or vacuum tweezers.
- Nonconductive, noncorrosive adhesive and toothpick or

other pointed tool for applying, or tape to anchor component for soldering.

- Table-mounted or handheld magnifier.

Work area

As with through-hole soldering, solder only in a well-ventilated room. Use a vent hood or fan to remove solder fumes from your work area. Do not eat or drink while soldering. Immediately after soldering, wash your hands to avoid ingesting lead. Illuminate your work area with a task lamp or other good lighting.

Procedure

As always, be sure your pc board artwork is clean before you solder. If the board has fingerprints, dirt, or other contamination, or if much time has passed since the board was etched, clean it with steel wool or another mild abrasive.

Tinning the artwork with immersion tin or solder, as described in chapter 8, is recommended for surface-mount circuits. The plating on each pad helps to create a good solder joint.

Soldering a chip resistor   The following section describes how to solder a chip resistor to a pc board using a soldering iron and wire solder.

1. Install a 0.002-inch screwdriver tip, or a similar type, in your soldering iron, and turn on the iron. To prevent thermal damage to components, keep the tip temperature below 700°F.
2. If you are using adhesive to hold the component on the board, use a toothpick or other pointed tool to apply a tiny dot of the adhesive to the underside of the body of the resistor (do not apply the adhesive to the terminations, which will be soldered), or apply the adhesive to the spot on the pc board where the resistor's body will rest.

   Use tweezers to pick up the resistor and carefully place it on its pads on the board. Press the resistor onto the board to ensure that its terminations are contacting the pads. Center the resistor on its pads.

   If you are using tape to anchor the resistor, place the edge of a piece of tape across the body of the resistor, leaving one termination uncovered. Carefully position the resistor on its pads and press the tape on the board to hold it in place.

3. Soldering a surface-mount component is much like soldering a through-hole component, except that the joint is smaller and thus can be made more quickly. Place the soldering iron's tip at the junction of the component's termination and its pad and apply solder (Fig. 9-21). When the solder melts, remove the solder and soldering iron from the joint.

4. After soldering, inspect the joint under a magnifier. You should see a smooth, shiny coating of solder that covers the component's termination and pad (Fig. 9-22). If necessary, remelt the joint with the soldering iron and add more solder, or use desoldering braid to remove any excess solder. If the component shifted in position during soldering, reheat the joint and reposition the component.

5. After soldering the first joint, you can remove the tape (if used). In the same way, solder the resistor's other termination to its pad.

Soldering other components Other surface-mount components solder in much the same way as chip resistors. Anchor the component on the board, carefully solder the first joint, inspect and adjust as necessary, solder the remaining joints, and inspect. The leads on surface-mount ICs have very close spacing. Take care to position the IC accurately on its pads, and inspect after

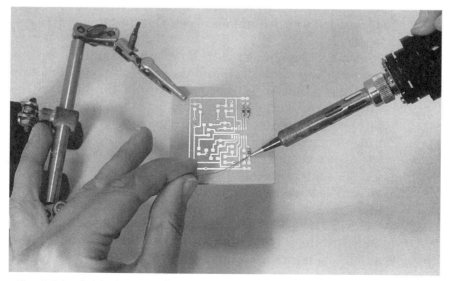

Fig. 9-21 *Soldering a surface-mount resistor.*

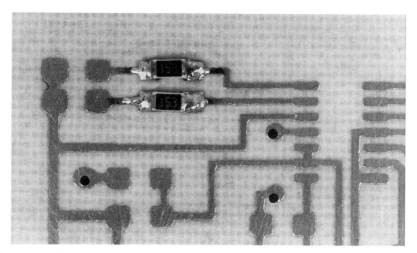

Fig. 9-22 *Surface-mount resistors hand-soldered to a pc board.*

soldering to ensure that there are no solder bridges connecting two or more leads.

Using solder paste

In addition to using traditional wire solder, you can solder surface-mount components with solder paste. A syringe, tooth-pick, or other pointed tool can be used to apply solder paste on pads or component leads, or you can dip the component leads directly into the solder paste. The sticky solder paste holds the component on the board, so there's no need for another adhesive to anchor the component.

A convenient way to apply solder paste is to use ESP's hand-held Dot-Maker dispenser, shown in Fig. 9-23. The Dot-Maker places a small dot of solder paste on each pad to be soldered. Caplettes of solder paste are available in a variety of solder pastes and fluxes.

To use the Dot-Maker, load a caplette of solder paste into the dispenser, place the tip of the caplette on the pad, and squeeze the dispenser's handle, causing a dot of solder paste to be deposited on the pad. Short squeezes result in small deposits and longer squeezes result in larger deposits. After applying solder paste to the pads, use tweezers to place the component on its pads.

On gull-wing ICs, instead of applying the paste to the indi-vidual pads, position the IC on its pads and lay a bead of solder

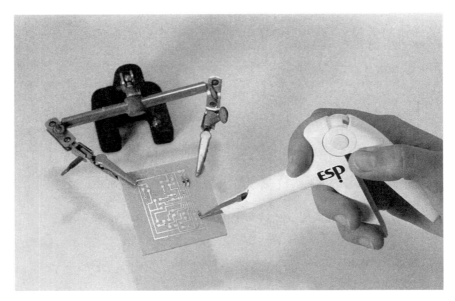

Fig. 9-23 *Applying solder paste with ESP's Dot-Maker.*

paste along each row of pads, on top of the leads. When the solder melts, it wicks onto the individual pads.

For small-volume soldering, you can melt solder paste with a soldering iron, hot-air source, or reflow hot plate. If you use a soldering iron with solder paste, be sure it's a low-wattage or temperature-controlled one. If you heat the joint too quickly, you might experience spattering, which occurs when solvents in the flux begin to boil and tiny balls of solder shoot across the board (Fig. 9-24). Preheating the joint first to around 200°F enables the flux solvents to evaporate and prevents spattering. Keep the tip temperature low and touch the soldering iron to only a corner of the pad. Experiment to find the best results with your iron.

When you melt the joints one by one with a soldering iron, the coating of solder paste must be thin. If the leads or terminations are floating on a thick layer of paste, as the first solder joint melts, one end of the component will drop onto the board, leaving the other end tilted on its unmelted paste. After the first solder joint sets, the component is fixed in place, and there's no easy way to move the other end into place.

One way to prevent this problem is to heat all of a component's joints at the same time. Special forked soldering tips are designed for this purpose and enable you to solder all of a component's connections at once. Tips are available to match different component packages.

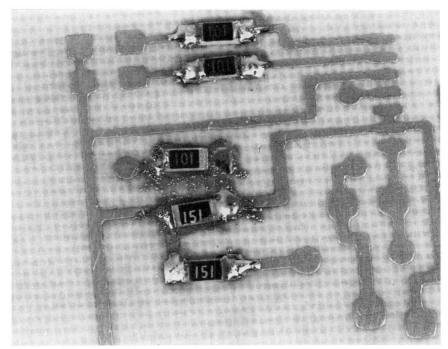

Fig. 9-24 *Solder balls caused by improper heating of solder paste.*

Instead of a soldering iron, you can use a hot-air source to melt the solder paste. A source with an adjustable temperature is best. Otherwise you regulate the temperature by moving the hot-air source toward or away from the joint as needed. Many industrial heat guns (of the type used to shrink heat-shrink tubing) have reducing baffles that direct a ¼-inch diameter source of hot air onto a component. Another possibility is to use a butane-fueled soldering iron with a hot-air attachment.

When using hot air to melt solder paste, be very careful not to overheat the component or base material. Preheat, then increase the temperature to melt the solder, and remove the hot-air source quickly after the joint forms. It takes practice to develop a good technique.

Another alternative for a heat source is a temperature-controlled reflow hot plate, such as OK Industries' SMT Reflow Hot Plate. After depositing solder paste and placing components on the board, you place the board on the hot plate, preheat, and then solder the board. When the solder paste reaches its melting temperature, the entire board solders at once.

To clean or not to clean

Circuits soldered with a no-clean flux do not require cleaning after soldering. This is a good reason to use solders with no-clean fluxes in your projects. Many boards soldered with noncorrosive, nonconductive rosin-based fluxes also don't require cleaning after soldering. Occasionally a board might require cleaning, either to remove flux residues that are corrosive or to remove sticky residues that might attract contaminants.

If you do clean a board, choose a flux remover that is compatible with your flux. Some flux removers are designed for use on rosin-based fluxes and nonpolar contaminants, while others are intended primarily for water-soluble fluxes and polar contaminants. For atmospheric ozone protection, look for a flux remover that uses an EPA-designated substitute for CFCs.

Most flux removers are available in spray cans with nozzles that direct the remover precisely where you want it. Use adequate ventilation and read and follow all safety precautions on the can.

To protect circuits that will operate in harsh environments, there are protective conformal coatings that you can apply to the assembled board by spraying, brushing, dipping, or flow-coating. The coatings protect against moisture, salt, fungus, corrosive vapors, and thermal and mechanical stress. The coatings are available in silicone, acrylic, urethane, and ultraviolet-curable epoxy formulations.

Mass-production soldering

Instead of hand-soldering each connection, commercial soldering of boards is often done in a single operation, using wave, vapor-phase, or infrared reflow soldering.

Wave soldering

A wave soldering machine generates a standing wave of molten solder. The pc board, with components inserted, is carried over the wave by a conveyor. In the process, solder is applied to all of the joints, and the entire board is soldered at once. Boards to be wave soldered usually have a solder mask that coats all areas of the board except the pads to be soldered.

Wave soldering works well for through-hole boards because one side of the board contains only solder pads and leads. With surface-mount boards, which have components and solder pads on the same side, it's harder to apply the solder wave evenly to all of the pads.

Vapor-phase soldering

In vapor-phase soldering, solder paste is applied by screen or stencil printing the paste onto the board, and components are pressed onto the solder paste. Soldering occurs as a liquid vaporizes and then condenses on the board's leads and pads. When the vapor condenses, it releases latent heat that solders the joints.

Infrared reflow soldering

In infrared reflow soldering, heating panels provide a controlled source of infrared heat to the board. As with vapor-phase soldering, the pads are coated with solder paste, which melts on heating.

Conductive adhesives

Solder isn't the only way to make an electrical connection to a pc board. Conductive adhesives are an alternative.

Step-by-step: Using conductive epoxy

Circuit Works' conductive epoxy is a two-part, quick-set, silver epoxy that you use much like other epoxies—mixing an epoxy and hardener and applying the mix to the surfaces to be attached. The difference is that this epoxy is conductive, with a volume resistivity of less than 0.001 ohm per centimeter. The epoxy bonds to many surfaces, including copper, solder alloys, epoxy laminate, many plastics, and glass.

Conductive epoxy can be used to glue surface-mount components to their pads, eliminating the need for solder. It's ideal for forming connections to heat-sensitive components. You can also use it to repair damaged pads or traces. For example, a dot of conductive epoxy will bridge a small opening in a trace.

When freshly applied, you can remove excess epoxy with a solvent. You can also rework or remove the epoxy with a hot soldering iron. But reworking is not as easy as with soldered connections, so the epoxy is best reserved for connections that you don't anticipate having to undo.

The following section describes how to use Circuit Works' conductive epoxy to fasten and connect a component to a pc board.

Materials

You will need the following materials:

- Circuit board with components to be soldered.
- Circuit Works 2400 Conductive Epoxy.
- Toothpick or other pointed applicator.

Work area

Work at a test bench or other worktable. Keep the epoxy away from open flames, use adequate ventilation, and avoid prolonged breathing of the vapors.

Procedure

You must use the conductive epoxy within 3 to 5 minutes of mixing the epoxy and hardener, so read through these instructions and gather your materials before you begin. Be sure any component leads, terminations, or traces to be epoxied are clean.

Open the two tubes and squeeze equal amounts on a scrap of paper or other disposable surface. Try to apply equal amounts of each, but an exact match isn't critical. Use the applicator or a toothpick to mix the epoxy and hardener.

When they are thoroughly mixed, use the applicator or toothpick to apply the mixture. For surface-mount components, apply a small amount to each pad on the pc board (Fig. 9-25). If possi-

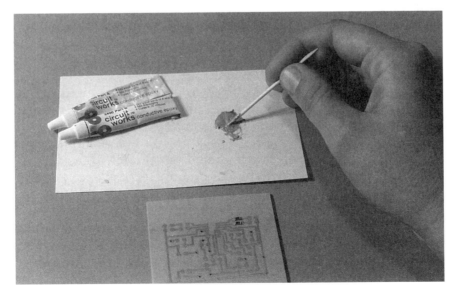

Fig. 9-25   *Using conductive epoxy to mount components.*

ble, also dab a small amount on each of the component's legs or terminations. Use a magnifier if necessary to view the surfaces. Be careful to avoid shorting pads or terminations together by using too much epoxy. After the epoxy is applied, press the component onto its pads. For through-hole components, install the component on the board before applying epoxy to the joints. You can remove excess epoxy with a solvent, preferably before the epoxy has cured.

After applying the epoxy, leave it to cure. The curing time depends on the temperature and the amount of material, with 10 minutes required to cure 1 ounce at 75°F. Smaller amounts and colder temperatures take longer to cure, up to 24 hours. To speed up curing, you can heat the epoxy to 100°F to 150°F for several minutes and allow it to cool. Be careful not to overheat the epoxy.

Conductive adhesive transfer tape

Another soldering alternative is 3M's Scotch 9703 Conductive Adhesive Transfer Tape. This unusual tape conducts only through the adhesive thickness, not along the plane of the tape. In other words, each strip of tape contains tiny conductors that connect the top and bottom surfaces of the tape, but do not connect to each other. You can use this tape to attach components that might be difficult or impossible to solder, such as flexible circuits or membrane switches.

Project enclosures

A project isn't complete until it is mounted in its enclosure and ready for use. The following sections summarize the steps involved in this final process.

Choosing an enclosure

Many types of enclosures are available for electronic projects. Popular enclosures include all types of plastic and metal boxes with removable covers. Figure 9-26 shows some examples. Some enclosures include slots into which you can slide a pc board of a matching size. Others require you to drill holes and add the mounting hardware for your board.

Metal enclosures are recommended where electrical shielding is desired or where the case must be grounded. Both aluminum and plastic are easy to drill and cut.

Fig. 9-26   *Examples of project enclosures.*

Some projects do not require a separate enclosure, such as computer boards intended for plugging into an existing motherboard or backplane. Sometimes you can use an unorthodox project case, such as a plastic package recycled from another product. If you're a woodworker, you can make your enclosures out of wood. Whatever enclosure you choose, be sure it can withstand the heat your circuits will generate, and any other environmental stresses (moisture, dust, accidental dropping, and so on) it might have to endure.

Board mounting methods

Three common methods of mounting a board in its enclosure are screws and standoffs, plugging the board into an edge connector, and sliding the board into slots on the enclosure. A board might use one or a combination of these. For example, many computer expansion boards use all three: they fit into a slot along one edge of the enclosure, the bottom plugs into an edge connector, and a bracket and screw fasten down one corner.

Screws and standoffs   To mount a board using screws and standoffs, mark and drill four mounting holes in the corners of the pc board. For large boards, you might want to drill additional holes midway between the corners for additional support. Drill corresponding holes in the enclosure.

To mount the board in the enclosure, you need a screw, nut, lock washer, and standoff for each hole. A standoff is a hollow cylinder that slides onto a screw and holds the pc board a uniform distance above the floor of the enclosure. For a board with trimmed component leads and no components mounted on the underside of the board, ¼-inch tall standoffs are high enough. Use screws about ⅜ inch longer than your standoffs to allow for the thickness of the board, enclosure, nut, and washer.

Edge connectors   If your board is designed to fit an existing edge connector, installation is easy. Carefully line up the connector fingers with the edge connector. Be sure the board is oriented correctly (not backwards). Push the board into its connector and examine the installed board to verify that the connector fingers line up with the contacts in the connector.

Slot mounting   Small boards can be cut to fit in a slot in an enclosure, and are installed by sliding the board into the matching slots. For this method, you must be sure to cut your board to the correct size: small enough to fit between its slots, but large enough not to slip out. Also be sure that no components are mounted near the edge of the board where they might interfere with the slots.

Cutting and drilling the enclosure

You can use a hand drill or drill press to cut holes in the enclosure. Holes might be needed for mounting the pc board, or for mounting indicators, controls, or connectors on the front and back panels of the enclosure. Carefully mark each hole's location in pencil before drilling. Use a center punch to indent the location before drilling, to help ensure accurate placement of the drill bit.

A tapered reamer will enlarge a small hole without requiring a large drill bit as shown in Fig. 9-27. To use the reamer, insert its tapered end into the hole to be enlarged as far as it will fit. Turn the reamer clockwise to cause its hardened-steel blades to cut and enlarge the hole to the desired size.

A nibbling tool will cut a rectangular hole from a round one (Fig. 9-28) that is ⅜ inch in diameter or larger. To use the nibbler, clamp its jaws along the edge of the hole to be enlarged and squeeze the handles to cut a rectangular piece from the edge of the hole. You can make a hole of any size by repeatedly nibbling small pieces. If necessary, you should file the edges of the final hole for a smooth surface.

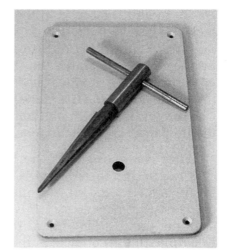

Fig. 9-27   *A tapered reamer will enlarge a drilled hole.*

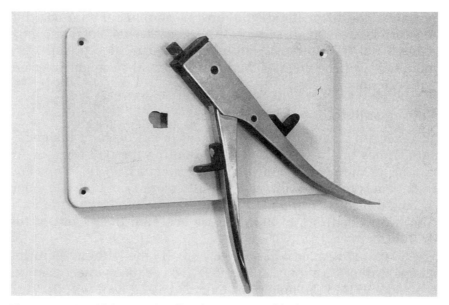

Fig. 9-28   *A nibbling tool will enlarge a round hole to a rectangular or odd-shaped hole.*

10
Repairing
pc boards

No matter how carefully you design and make your pc boards, eventually you will find yourself with a board in need of repair. Reasons for repair include installation of incorrect components, correct components installed incorrectly, failed components, solder bridges, cold solder joints, mistakes in board layout, and design modifications. This chapter will focus on how to make neat, effective repairs and modifications to pc boards.

Replacing components

Many circuit repairs involve removing a soldered component from a board. An IC might have 40 or more legs, and removing an IC requires melting the solder joint at each leg. The time and effort required to remove a soldered IC is a good argument for using a socket for easy removal and replacement of ICs. Still there are times when sockets are impractical or impossible, as with many surface-mount ICs.

When removing a soldered component, you can choose between destructive and nondestructive methods. Destructive removal is usually quicker, but it destroys the IC in the process. Nondestructive methods require more care, but allow you to salvage the component for reuse. With both methods, you can make use of a solder extractor or desoldering braid, or both.

Using a solder extractor

A solder extractor is a handheld tool that generates a vacuum that sucks molten solder from a heated joint. Several varieties of

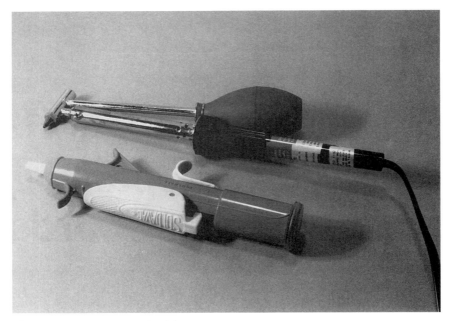

Fig. 10-1 *Two types of solder extractors.*

solder extractor are available. Figure 10-1 shows two low-cost types. One has a heat-resistant hollow tip and a spring-loaded plunger, while the other combines a heated tip with a squeeze bulb that creates the suction.

To use a spring-loaded solder extractor, set the plunger by pushing down on it. Then, holding a heated soldering iron in one hand and the solder extractor in the other, apply the soldering iron to melt the solder joint to be desoldered. When the solder melts, press the tip of the solder extractor against the joint and release the plunger. The resulting vacuum sucks the molten solder into the body of the extractor. To use an extractor that includes a heated tip, you must plug in the extractor to preheat it, then press the heated, hollow tip on the joint to be desoldered, and squeeze and release the bulb to suck the solder into the tube.

A heat-resistant squeezable bulb is another very simple form of solder extractor for use with a soldering iron. Some higher-priced extractors contain a vacuum pump or connect to an outside source of compressed air for fast desoldering with precise control. For desoldering tiny surface-mount joints, use a miniature model.

Step-by-step: Nondestructive component removal

You can nondestructively remove most leaded components, both through-hole and surface-mount, with a solder extractor and the following procedure.

Materials

You will need the following materials:

- The pc board with the component to be removed.
- Soldering iron.
- Solder extractor.
- Needle-nose pliers.

Work area

The environment for repairing pc boards is the same as for soldering. Be sure to have adequate ventilation, use good lighting, and wash your hands after soldering.

Procedure

Follow these simple steps:

1. Turn on your soldering iron and set the plunger, if any, on the extractor.
2. Place the tip of your heated iron on the first solder joint to be removed.
3. When the solder melts, press the tip of the solder extractor against the solder joint, keeping the soldering iron tip on the joint if possible. Release the plunger to draw the molten solder into the extractor. Remove the soldering iron and extractor from the joint.
4. Examine the desoldered joint. With a sharp probe or needle-nose pliers, try to free the lead from its pad or hole. If the lead does not move freely, repeat the solder removal process. The solder extractor works best on intact joints. Bits of remaining solder can be hard to remove from partially desoldered joints. If the joint doesn't desolder, resolder the joint and try again.
5. When the first lead is free, repeat the procedure on the other component leads. Figure 10-2 shows the solder joints of a partially desoldered IC. With through-hole

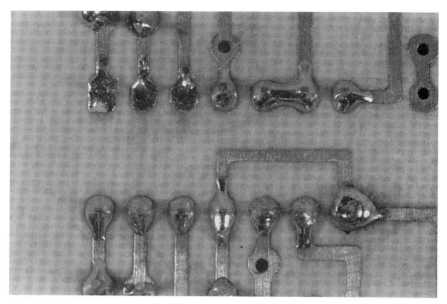

Fig. 10-2   *Three of the solder joints on this 14-pin IC have been desoldered using a solder extractor.*

Fig. 10-3   *Grip and twist to remove a desoldered component that is attached to a pc board by adhesive.*

components, when all of the joints have been desoldered, you should be able to lift the component from the board. If only one lead remains soldered to its pad, you can work it free by gently wiggling the lead with needle-nose pliers as you heat the joint with the soldering iron.

If a surface-mount component has been attached with adhesive, you need to break the adhesive bond to remove the component from the board. Grasp the component body with needle-nose pliers or another tool and twist to pull the component off the board (Fig. 10-3).

Step-by-step: Destructive component removal

Destructive removal is easier than nondestructive removal and might make more sense for low-cost components.

Materials

You will need the following materials:

- The pc board with the component to be removed.
- Soldering iron.
- Solder extractor.
- Wire cutters.

Procedure

Follow these simple steps:

1. Turn on your soldering iron.
2. Use wire cutters to clip off each of the component's leads. When all leads have been cut, remove the component body from the board, leaving the leads attached to their pads (Fig. 10-4).
3. Grip one of the leads with needle-nose pliers. Apply the tip of a hot soldering iron to the joint. When the solder melts, pull the lead out of its hole.
4. Repeat for the other leads.
5. When all the leads have been removed, use a solder extractor or desoldering braid to remove excess solder from the pads.

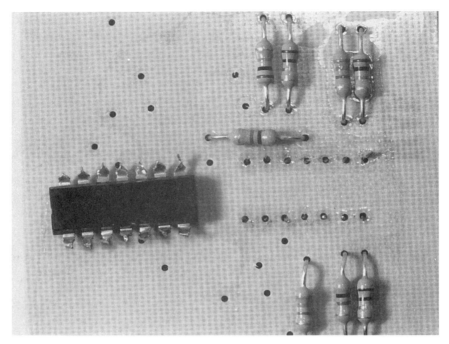

Fig. 10-4 *Destructive removal involves removing a component by clipping the leads from the body, then desoldering and removing each lead individually.*

Using desoldering braid

Desoldering braid is an alternative or supplement to the solder extractor. The braid is made of thin strands of copper woven into a single, flat band. Flux is included to promote solderability. To use desoldering braid, press it onto the joint to be desoldered and place a heated soldering iron tip on the braid (Fig. 10-5). When the solder joint melts, the molten solder wicks up into the braid.

Desoldering braids are available in varying widths from 0.025 inch to 0.210 inch, for use on different-size solder joints. For best results, match the width of the braid to the width of the pad to be desoldered.

Hot-air component removal

Hot-air removal is the complement to hot-air soldering. Both enable you to melt all of a component's solder joints at once. Special hot-air rework stations are designed for hot-air component removal. Lower-cost options include a hot-air gun with a reduc-

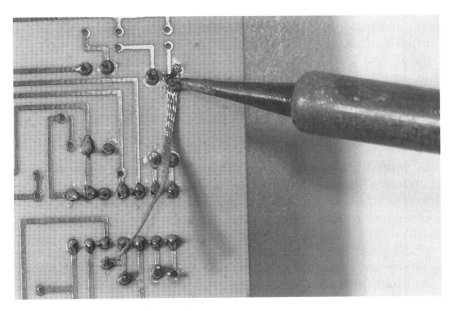

Fig. 10-5   *Using desoldering braid.*

ing baffle, a butane-fueled soldering iron with a hot-air attachment, or any hot-air source that reaches soldering temperatures and can be directed at a single component. The air must be hot enough to melt the solder, but not so hot that the pc board's base material or the components are damaged by the heat.

Hot-air removal is especially useful for J-leaded surface-mount components. Because the solder joints on these components are underneath the package, it's hard to press a solder extractor or desoldering braid onto them.

To desolder with hot air, preheat the component by setting the hot-air source to 200°F or by holding the hot-air source 1 inch or more from the component. Don't allow the solder and flux to melt.

After preheating, increase the hot-air temperature to melt the solder. Desoldering should occur within a few seconds after increasing the temperature. Be careful not to overheat. Remember that components attached with adhesive need to be twisted to break the adhesive bond.

Installing replacement components

To install a new component where one was removed, first inspect the pads. Be sure all through-hole pads have an open lead hole in the center. Be sure all surface-mount pads are coated with a thin, even layer of solder, so the component will lie flat on the pads. Use a solder extractor or desoldering braid to open any plugged holes or to remove excess solder from a pad. Solder the replacement component using the techniques described in chapter 9.

Repairing pads and traces

Occasionally a pad or trace will tear or lift from a board's base material. This can be caused by too much heat during soldering, too much heat or suction during removal of a component, physical damage to the board, or poor adhesion of the pad or trace to begin with.

If a pad lifts from the board, you can use epoxy cement to glue it back. Mix tiny, equal amounts of epoxy and hardener and use a toothpick or other pointed tool to apply a small amount of the mixed epoxy at the pad's location on the board. Press the lifted pad back into place and leave it undisturbed to harden.

If a pad has been destroyed, you can solder a jumper wire to substitute for the connection that should have been soldered to the pad. Cut a length of insulated wire slightly longer than the trace it will substitute for. For most signal traces, AWG #30 wire is sufficient. For traces that carry more than 100 milliamperes of current, use a thicker, lower-gauge wire. Use solid wire, not stranded, for easier soldering.

Strip ¼ inch of insulation from each end of the wire. Wrap one end of the wire around the component leg at the missing pad, and wrap the other end around the component leg to which the missing pad connected. Solder both connections and trim the wire ends if necessary. This technique works for both through-hole and surface-mount boards, though the spacing is much tighter for jumpers soldered to surface-mount components.

In a similar way, you can repair an open trace or add a connection by soldering a jumper wire between component pins to replace the broken connection as shown in Fig. 10-6.

If a jumper wire is long, it's a good idea to tack it onto the board every inch or so to hold it in place. A dot of silicone rubber sealant or nonconductive epoxy will serve for this.

To bridge a tiny open area in a trace, you can solder a short length of wire directly onto the trace. Be sure the traces to be soldered to are clean. Scrape lightly with an artist knife if necessary. Strip ½ inch or so from a length of insulated wire. Holding the wire with needle-nose pliers, place the stripped end across

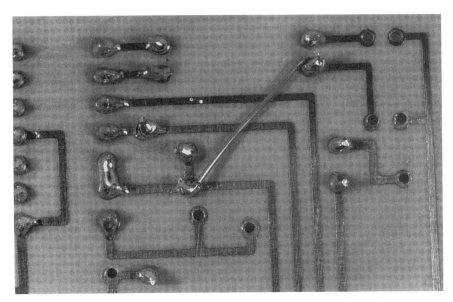

Fig. 10-6   *You can add a connection by soldering an insulated jumper wire between component pins.*

the gap to be bridged, and solder the wire on both sides of the gap to make the connection. Clip the wire at the joint.

If you do a lot of board repair, you might want to invest in a repair kit such as one of PACE's CIR-KITs (Fig. 10-7). The kits include pads and traces that you can apply to a board, eyelets/funnelets for repairing plated through-holes, bonding materials, and the necessary tools for using these items.

Circuit modifications

No matter how carefully you design your pc board artwork, there will be times when you find that your pc board's artwork doesn't exactly match your circuit's schematic. Maybe you forgot to include a trace, or accidentally misrouted a connection, or forgot to include a component in the original artwork. Another reason for circuit modifications is when circuit testing reveals that the original circuit design was flawed, and that connections or components need to be changed to get the circuit to work or to improve its performance. Minor modifications to pc board artwork are easily made. You can add connections with jumpers, using the techniques described previously.

If you need to break a connection, cut the trace with an artist knife. Make two cuts through the copper about ⅛ inch apart (Fig. 10-8). Then apply the tip of a hot soldering iron to the segment of

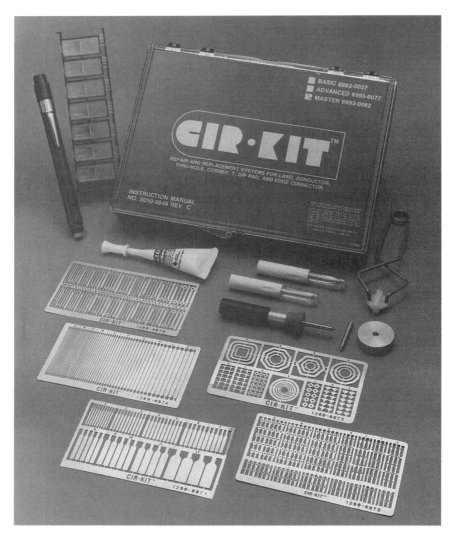

Fig. 10-7 *A repair kit for pc boards. (PACE, Inc.)*

the trace between the cuts, and push the cut segment off of the board. The result is an open circuit between the remaining halves of the trace (Fig. 10-9).

You can add components to a pc board without scrapping the board and starting over. For example, you can add a through-hole resistor or capacitor by wrapping the leads around other component leads to which they connect and soldering, as shown in Fig. 10-10. Place short lengths of heat-shrink tubing over the component's leads to prevent them from shorting to other components.

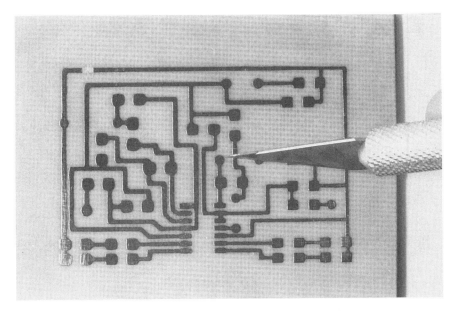

Fig. 10-8   *Cutting a trace.*

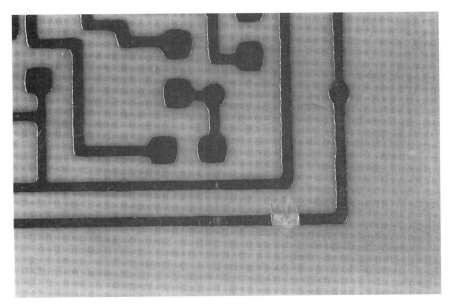

Fig. 10-9   *To open a trace, make two cuts and remove the section of trace between them.*

If a circuit requires major modifications, such as adding ICs or more than a few discrete components, you're probably better off starting over with fresh artwork and a new pc board.

Fig. 10-10   *Modifying a circuit by adding a resistor.*

Step-by-step:
Using conductive ink

Conductive ink is another way to repair pads and traces. Circuit Works' conductive pen is filled with a specially formulated silver-bearing polymer, which you can use to draw traces, pads, shielding, or other conductive lines or areas. The pen will draw on both rigid and flexible circuit board base materials.

Resistivity of the ink measured across a square of any size, applied at a thickness of 0.001 inch, is 0.03 to 0.05 ohm after heat curing. Normal trace width is 0.05 inch, so the pen is suitable for many applications, although the traces are too wide for drawing connections to surface-mount ICs. The pen contains enough ink to draw almost 100 feet, and has a shelf life of 6 months after first use.

Bottled conductive inks are also available, which can be brushed on or applied with a toothpick. In a pinch, you can even use the conductive ink sold in hardware stores and auto repair shops for repairing car windshield defrosters. Circuit Works' conductive adhesive, described in chapter 9, is another useful product for repairing pc board artwork.

The following section describes how to use conductive ink to draw artwork on a pc board.

Materials

You will need the following materials:

- Circuit board requiring addition of a trace or pad.
- Circuit Works 2200 Conductive Pen.
- Hot-air gun or other heat source for heat curing (optional).

Work area

Work at a test bench or other worktable with good lighting. Keep the pen and ink away from open flames, and have adequate ventilation.

Procedure

Follow these simple steps:

1. To use the pen, shake it vigorously for 20 to 30 seconds to ensure that the conductive material is mixed.
2. Begin by drawing some practice lines to get a feel for how to use the pen. To draw, press the tip lightly on the drawing surface and squeeze the pen's flexible barrel firmly between your thumb and index finger. The pen's spring-loaded tip allows the conductive ink to flow smoothly from the pen onto the drawing surface. To regulate the flow rate, adjust the pressure on the barrel. It might take some practice before you are drawing smooth, even lines of the thickness you desire.
3. When you're ready to draw onto your circuit board, proceed in the same way to draw your pads or traces (Fig. 10-11). Wipe the pen tip after use to prevent clogging.
4. The ink dries in 3 to 5 minutes, and reaches its maximum air-dried conductivity after 20 to 30 minutes.
5. Heat curing is recommended for maximum conductivity and to create a termination that can be soldered. Cure the ink at 250°F to 300°F for 3 to 5 minutes to increase conductivity, and for 10 to 15 minutes to create a solderable termination. If you use a hot-air gun, be careful not to overheat and burn or damage the pc board's base material.
6. To solder a cured termination, use the lowest temperature possible to melt the solder, and apply heat for less than 5 seconds. Special low-temperature solder is recommended.

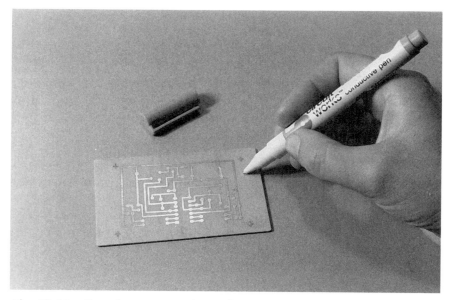

Fig. 10-11 *Drawing traces with conductive ink.*

11
Projects

This chapter presents three projects from schematic to soldered, ready-to-use pc board:

- Project 1: Five-volt power supply.
- Project 2: All-purpose pulser/flasher.
- Project 3: Two-channel logic probe.

The projects illustrate a variety of circuit types, including high- and low-power circuits, and through-hole and surface-mount devices. The projects are intended as examples to get you started in making your own pc boards. You can use these, or similar simple projects, to experiment with the techniques and procedures described in this book. From here, you can move on to designing pc boards for more complex circuits, including custom designs for projects of your own.

Project 1: Five-volt power supply

Figure 11-1 shows a general-purpose, 5-volt, 1-ampere power supply. The supply converts 115-volt ac power into a stable, regulated 5-volts dc. You can use this supply to power many circuits that require a 5-volt supply.

About the circuit

Figure 11-2 is the schematic diagram for the 5-volt supply. The circuit consists of three main elements: a transformer, a rectifier, and a regulator.

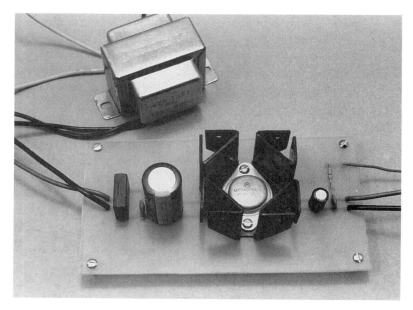

Fig. 11-1   *Five-volt power supply.*

Transformer T1 converts 115-volt power from a wall socket to 12.6-volts ac. The transformer also electrically isolates the power supply circuits from the ac power line.

BR1 is a bridge rectifier that converts T1's ac output current into a pulsed dc current.

U1 is a 7805, 5-volt regulator whose output is a steady, regulated 5-volts dc.

Capacitors C1 and C2 are filter capacitors that smooth BR1's output and ensure that U1's input is always at least 7.5-volts dc, the minimum input required for proper operation. Capacitors C3 and C4 help to ensure that U1's output is a steady 5 volts in spite of load variations.

LED1 is a power-on indicator, and resistor R1 limits current through LED1.

About the artwork

Figure 11-3 is actual-size pc board artwork for the power supply. Figure 11-4 is a parts placement diagram. Table 11-1 is a parts list for the project. The artwork is shown in both actual and mirror-image orientations to accommodate different methods of image transfer.

Because the circuit traces might carry over 1 ampere, the artwork uses thick, 0.06-inch wide traces. The wide traces cause no

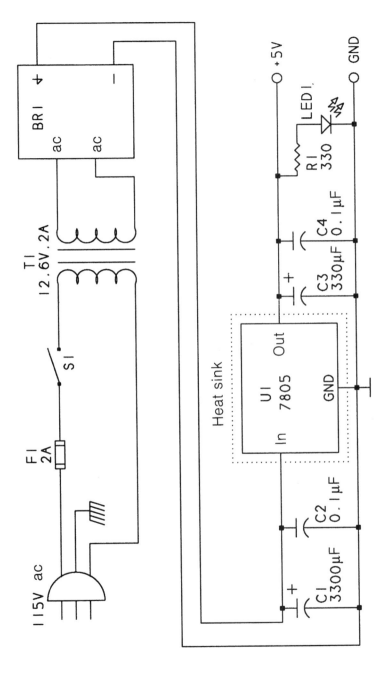

Fig. 11-2 *Schematic diagram for the 5-volt power supply.*

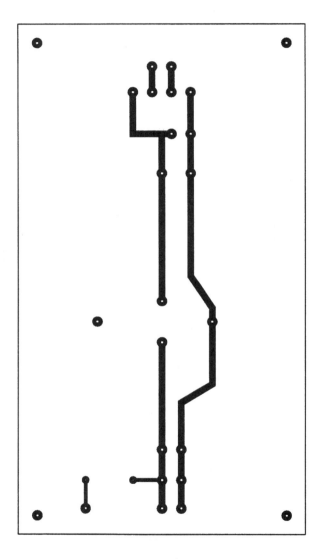

Bottom-side orientation

Fig. 11-3 (a) *Actual-size pc board artwork for the 5-volt supply.*

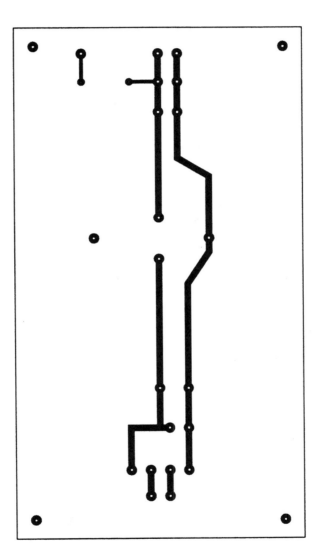

Mirror image

Fig. 11-3 (b)   *Actual-size pc board artwork for the 5-volt supply.*

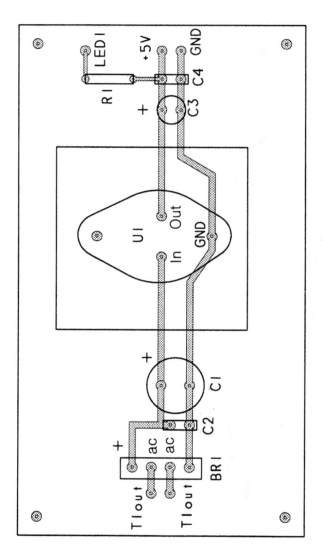

Fig. 11-4 *Parts placement diagram for the 5-volt supply.*

Table 11-1   Parts list for 5-volt power supply.

Resistors
R1	330 Ω, ¼-W, 10% tolerance

Capacitors
C1	3300-μF, 35 WVdc, aluminum electrolytic
C2, C4	0.1-μF, 35 WVdc, ceramic disc
C3	330-μF, 16 WVdc, aluminum electrolytic

Semiconductors
BR1	Bridge rectifier, 2-amp
LED1	Light-emitting diode, red
U1	7805 +5-volt regulator in TO-3 package

Other components
F1	Fuse, 2-amp, slow-blow
S1	SPST switch, normally open, toggle, panel mount
T1	Transformer, 115/12.6-volt ac, 2-amp

Miscellaneous
Banana jacks or binding posts (2), panel mount
115-volt ac line cord and plug
Plastic or metal enclosure
Heat sink for TO-3 package, 10 watts or greater
Heat sink compound
Mounting hardware
Heat-shrink tubing
Panel-mount hardware for LED
In-line fuse holder
Rubber grommet for ac line cord

problems in laying out the artwork because the circuit components have widely spaced leads.

To prevent U1 from overheating, it should be mounted on a heat sink. U1's input is 12.6-volts RMS, and its output is 5-volts dc. Energy that is not converted into output current is dissipated as heat by U1. A large area is provided on the board for the heat sink.

The transformer leads are soldered along one edge of the board, and the power supply output leads are along the opposite edge. The transformer is mounted off-board because it is a large, heavy component and its connections are easily wired by hand. Fuse F1, switch S1, and LED1 also mount off-board.

Project construction

Follow these simple steps:

1. To make the power supply, gather the components in the parts list.

Capacitors C1 and C3 are radial-lead devices, but if you wish, you can substitute axial-lead components and modify the artwork to reflect this.

Be sure the pc board and transformer will fit in your enclosure. Modify the artwork to rearrange the components if necessary.

2. Using the provided artwork or your own design, make, etch, and drill a pc board using any of the methods described in this book. Be sure your holes are large enough for the thick leads of U1 and BR1. Don't forget to drill the mounting holes for the board and the heat sink. Enlarge the holes as necessary to match your mounting hardware.

3. Install and solder R1, C1 through C4, and BR1. Be sure to observe the correct orientations for BR1, C1, and C3.

4. Squeeze a small amount of heat sink compound onto the bottom of U1 and use a toothpick to spread the compound over the bottom surface. Position the heat sink on the board so that its holes line up with U1's holes on the board, and install U1 onto the heat sink and board. Use two screws, nuts, and washers to fasten U1's case onto the board. Solder U1's leads to the board.

5. Use insulated hookup wire of AWG #22 or lower for all wiring. Insulate all connections between T1 and the 115-volt line cord with heat-shrink tubing. T1 should have four leads—two connecting to its 115-volt winding, and two connecting to its 12.6-volt winding. If there is a center-tap lead, cut it off, because it's unused in this project.

Solder the two wires of $T1_{out}$ to T1's 12.6-volt leads. Using the schematic diagram as a reference, wire S1 and F1's in-line fuse holder. Don't solder the 115-volt line cord to the circuit yet. Solder wires to +5V, GND, and LED1 on the pc board.

6. Prepare the enclosure by drilling mounting holes for the pc board and T1 in the floor of the enclosure. On the front panel, drill holes for LED1, S1, and the banana jacks to +5V and GND. On the rear panel, drill a hole for the line cord's rubber grommet.

7. Install the pc board and transformer in the enclosure. Thread the wires connecting to +5V, GND, and LED1 through their matching holes. Solder the +5V and GND wires to banana jacks. Solder LED1's wire to the LED's anode, and solder a second wire from the LED's cathode to GND at the banana jack.

8. Install the grommet and thread the 115-volt line cord through its hole. Inside the enclosure, tie the cord in a loose knot to act as a strain relief. Using heat-shrink tubing to insulate, solder the connections to the line cord as shown in the schematic diagram. In the line cord, the "hot" wire (black in standard U.S. color code) connects to F1 and the neutral wire (white) connects to the opposite side of T1's 115-volt winding. In a metal enclosure, you can ground the case by connecting the line cord's safety ground (green) to the enclosure at one of the mounting screws. Do not wire the safety ground to any of the power supply circuitry.

9. Install the banana jacks, S1, and LED1, and insert F1 in its holder.

Checkout and use

Double check to be sure that all 115-volt connections are insulated and correctly wired. Do not power up the project until you are confident that the wiring is correct.

With S1 off (open), plug the line cord into a 115-volt wall socket. Close S1 and measure with a voltmeter from +5V to GND. It should read between 4.8 and 5.2 volts. To test the supply at 1 ampere, connect a 5-ohm, 10-watt resistor between +5V and GND and measure the output voltage as before to ensure that it is still within specifications.

Enhancements and modifications

You can use this project as the basis for many similar power supply projects. A 12- or 15-volt supply using a 7812 or 7815 regulator would be similar, but would require a transformer with a higher output voltage. A variable supply designed around an LM317HV would require adding a potentiometer and possibly a meter on the front panel for setting and viewing the output.

Whatever the design, be sure your pc board traces are wide enough to carry the current required, and that heat-sinking keeps the circuits from overheating.

Project 2: All-purpose pulser/flasher

Project 2 is an all-purpose pulser/flasher designed around a 555 timer. The pulser's output is a rectangular wave whose "high" and "low" times vary according to the component values used. You can build the circuit using either through-hole or surface-

Fig. 11-5   All-purpose pulser/flasher, built with through-hole components.

Fig. 11-6   All-purpose pulser/flasher, built with surface-mount components.

mount components, as shown in Figs. 11-5 and 11-6. Both versions are described later.

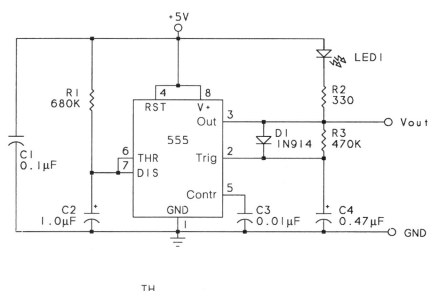

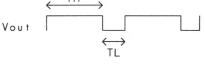

$$TH = 1.1(R1)(C2) = 0.75 \text{ SECOND}$$

$$TL = 1.1(R3)(C4) = 0.25 \text{ SECOND}$$

Fig. 11-7   *Schematic diagram for the pulser/flasher.*

About the circuit

Figure 11-7 is the circuit schematic. U1 is a 555 timer, which can be a bipolar or CMOS type. U1 outputs a rectangular wave at a desired frequency at pin 3. The values of R1 and C2 determine the output's high period, and the values of R3 and C4 determine the output's low period, as described by the formulas in the schematic. When pin 3 is low, LED1 is on.

You can change the output's frequency and duty cycle to suit your application by substituting different values for R1, R3, C2, and C4.

V_{out} can be used as a general-purpose rectangular-wave output. The circuit can be powered by any 5-volt supply or by four 1.5-volt batteries in series.

About the artwork

Both through-hole and surface-mount artworks are included in the following figures:

- Figure 11-8—actual-size artwork, through-hole version, bottom side and mirror image.
- Figure 11-9—parts placement diagram, through-hole version.
- Table 11-2—parts list, through-hole version.
- Figure 11-10—actual-size artwork, surface-mount version, top side and mirror image.
- Figure 11-11—identical to Fig. 11-10, but at ×2 scale for greater clarity.

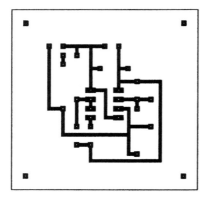 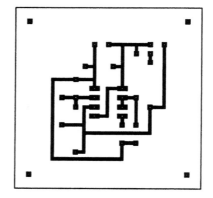

Bottom-side orientation   Mirror image

Fig. 11-8   *Actual-size pc board artwork for the pulser/flasher (through-hole version).*

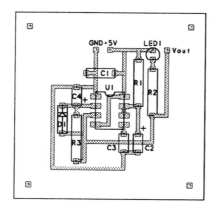

Fig. 11-9   *Parts placement diagram for the pulser/flasher (through-hole version).*

**Table 11-2 Parts list for
logic pulser, through-hole version.**

Resistors (¼-W, 5% tolerance)
R1	680,000 Ω
R2	330 Ω
R3	470,000 Ω

Capacitors (16 WVdc)
C1	0.1-μF, ceramic disc
C2	1.0-μF, tantalum electrolytic
C3	0.01-μF, ceramic disc
C4	0.47-μF, tantalum electrolytic

Semiconductors
D1	1N914 silicon signal diode
LED1	Light-emitting diode
U1	555 timer

Miscellaneous
Mounting hardware
Enclosure
IC socket, 8-pin
Rubber grommet

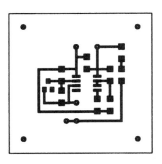 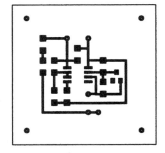

Top-side orientation   Mirror image

Fig. 11-10 *Actual-size pc board artwork for the pulser/flasher (surface-
mount version).*

- Figure 11-12—parts placement diagram, surface-mount
  version.
- Table 11-3—parts list, surface-mount version.

Notice that the through-hole artwork is about 50 percent
larger than the surface-mount artwork. On the surface-mount
version, jumper J1 is included so that a narrow trace does not
have to be routed underneath C2 and C3. On the surface-mount
board, diode D1 is in a three-lead TO-23 package, with one lead
unused.

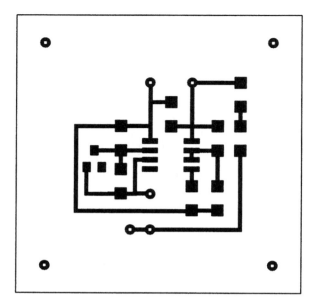

Top-side orientation

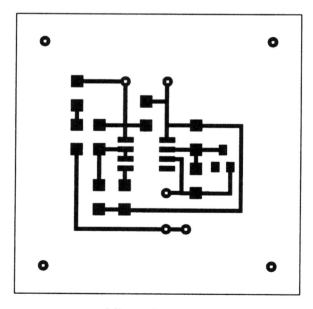

Mirror image

Fig. 11-11 *The pc board artwork (×2 scale) for the pulser/flasher (surface-mount version).*

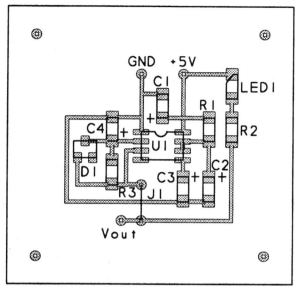

Fig. 11-12   *Parts placement diagram for the pulser/flasher (surface-mount version).*

**Table 11-3   Parts list for
logic pulser, surface-mount version.**

Resistors (⅛-W, 5% tolerance, 1206 size)
R1	680,000 Ω
R2	330 Ω
R3	470,000 Ω

Capacitors (16 WVdc, 1206 size)
C1	0.1-µF, tantalum
C2	1.0-µF, tantalum
C3	0.01-µF, tantalum
C4	0.47-µF, tantalum

Semiconductors
D1	Silicon signal diode, TO-23 package
LED1	Light-emitting diode, surface mount
U1	555 timer, SO-8 package

Miscellaneous
Mounting hardware
Enclosure
Rubber grommet

Because the artwork is small, you can easily make multiple copies of the pc board by transferring several artwork images onto a single board and cutting them apart after etching.

Construction: Through-hole version

Follow these simple steps:

1. Make and etch the pc board using any of the methods described in this book. Drill component and mounting holes.
2. Install and solder the components in this order: U1's socket, R1 through R3, C1 through C4, and D1. Be sure to observe correct polarities for C2, C4, D1, and LED1, as indicated in the parts placement diagram. Choose resistor and capacitor values to match the output frequency and duty cycle you desire.
3. If you are going to mount the board in an enclosure, you'll want to mount LED1 and jacks for the 5V and GND inputs and the V_{out} and GND outputs on the enclosure, and wire these to the pc board.

   Strip the ends and solder lengths of insulated wire to the GND and +5V pads. To mount V_{out} and LED1 on the front panel, also strip and solder lengths of insulated wire to their pads.
4. Drill mounting holes for the pc board in the floor of the enclosure. In the front panel, drill holes for LED1 and the jacks. Mount the pc board in its enclosure. Solder or otherwise connect the wires from +5V, V_{out}, and LED1 to their matching jacks or components. Install U1 in its socket.

Construction: Surface-mount version

You can also build the pulser/flasher using surface-mount components. Build the surface-mount version in the same way as the through-hole version, with these differences:

- U1 does not use a socket, but solders directly to the pc board.
- Observe correct polarities for C1 through C4 and LED1 as indicated in the parts placement diagram.
- Remember to drill through-holes for board mounting J1, V_{out}, GND, and +5V. To mount LED1 off-board, also drill a hole at the pad for its cathode.
- Remember to use appropriate techniques for placing and soldering the tiny surface-mount devices.

Checkout and use

To test the pulser, connect a 5-volt power supply or four 1.5-volt batteries in series to +5V and GND, and observe or measure the output at pin 3. LED1 should flash whenever pin 3 is low. Using the component values in the schematic, LED1 flashes on for 0.25 second, then off for 0.75 second.

Enhancements and modifications

For a variable pulser, install potentiometers at R1 and R3. For panel-mount adjustments, wire the potentiometers to the pads of R1 and R3 without altering the pc board artwork. For board-mounted potentiometers, revise the artwork to match the lead configurations of the potentiometers.

Project 3: Dual logic probe

Project 3 is a dual logic probe that you can use in testing and troubleshooting digital circuits. Figure 11-13 compares the project built with through-hole and surface-mount components.

About the circuit

Figure 11-14 is a schematic diagram for the dual logic probe. The circuit can monitor two digital signals. To use the probe, connect

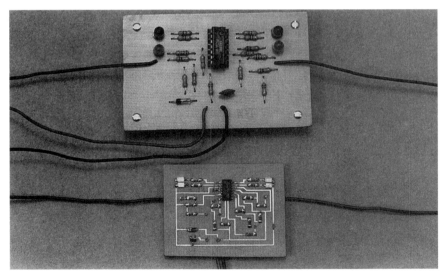

Fig. 11-13 *Comparison of the dual logic probe built with through-hole (top) and surface mount (bottom) components.*

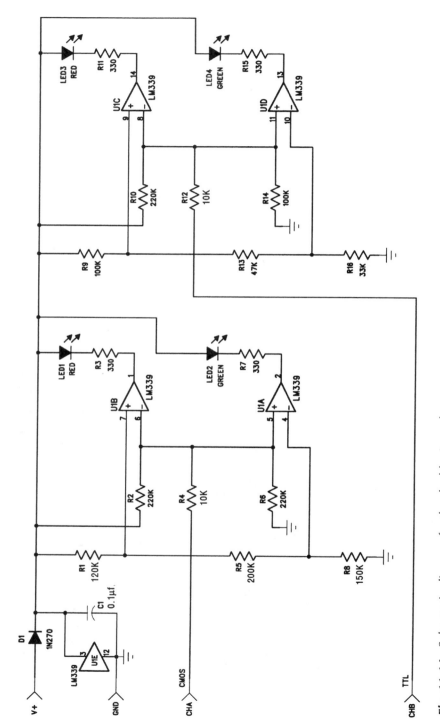

Fig. 11-14 *Schematic diagram for the dual logic probe.*

V+ and GND to the power supply and ground in your circuit, and connect CHA (channel A), CHB (channel B), or both to the signal or signals you want to monitor.

If the signal at CHA is a logic high, red LED1 lights; if the signal is a logic low, green LED2 lights. If the signal being monitored is not at a valid logic level, neither LED is on. In the same way, LED3 and LED4 show the logic state of the signal at CHB. CHA is configured to respond to valid CMOS logic levels, and CHB is configured to respond to valid TTL logic levels.

The circuit consists of two nearly identical halves, each using two of the four comparators of an LM339. In the first half, resistors R1, R5, and R8 are a voltage divider that determines the valid logic levels. As shown, a valid logic high must be at least 3.5 volts, and a valid logic low must be equal to or less than 1.5 volts. These levels are standard for CMOS circuits powered at 5 volts.

If the voltage at CHA is greater than 3.5 volts, pin 1 at U1B goes low and LED1 turns on. If CHA is less than 1.5 volts, pin 2 at U1A goes low and LED2 turns on. If the voltage at CHA is between 1.5 and 3.5 volts, neither LED is on.

Resistors R2 and R6 ensure that the LEDs are off if CHA is not connected to a signal. Resistor R4 limits current to protect U1 in case CHA accidentally connects to a negative voltage.

Resistors R9 through R16, U1C, U1D, LED3, and LED4 form a second logic probe for CHB. In this half, R9, R13, and R16 are chosen to correspond to valid TTL high and low voltages. LED3 lights red when CHB is greater than 2 volts, and LED4 lights green when CHB is less than 0.8 volt.

Diode D1 protects the circuits by blocking current flow in case the V and GND probes are accidentally reversed, and capacitor C1 provides power supply decoupling.

About the artwork

Both through-hole and surface-mount artworks are included in the following figures:

- Figure 11-15—actual-size artwork, through-hole version, bottom-side orientation.
- Figure 11-16—actual-size artwork, through-hole version, mirror image (component-side orientation).
- Figure 11-17—parts placement diagram, through-hole version.

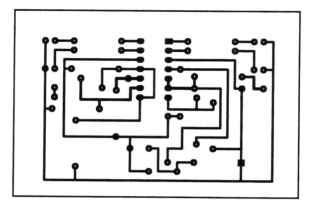

Fig. 11-15 *Actual-size pc board artwork for the logic probe, through-hole version, bottom (solder-side) orientation.*

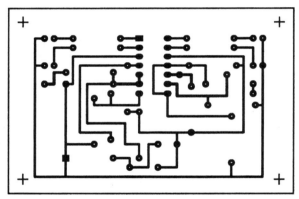

Fig. 11-16 *Actual-size pc board artwork for the logic probe, through-hole version, mirror image, top (component-side) orientation.*

- Table 11-4—parts list, through-hole version.
- Figure 11-18—actual-size artwork, surface-mount version, top (component) side.
- Figure 11-19—actual-size artwork, surface-mount version, mirror image.
- Figure 11-20—identical to Fig. 11-18, but at ×2 scale for greater clarity.
- Figure 11-21—identical to Fig. 11-19, but at ×2 scale for greater clarity.
- Figure 11-22—parts placement diagram, surface-mount version.
- Table 11-5—parts list, surface-mount version.

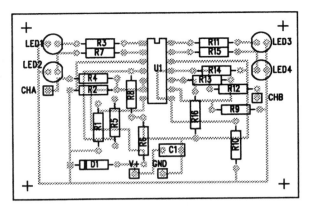

Fig. 11-17 *Parts placement diagram for the logic probe, through-hole version.*

**Table 11-4   Parts list for
logic probe, through-hole version.**

Resistors (¼ -W, 5% tolerance)

R1	120,000 Ω
R2, R6, R10	220,000 Ω
R3, R7, R11, R15	330 Ω
R4, R12	10,000 Ω
R5	200,000 Ω
R8	150,000 Ω
R9, R14	100,000 Ω
R13	47,000 Ω
R16	33,000 Ω

Capacitors (16 WVdc)

C1	0.1-µF ceramic disc

Semiconductors

D1	1N270 or similar germanium diode
LED1, LED3	Light-emitting diode, red
LED2, LED4	Light-emitting diode, green
U1	LM339 quad comparator

Miscellaneous

Test leads with alligator or mini probe tip on one end,
   banana plug on the other (4)
Banana jacks (4)
IC socket, 14-pin

The through-hole artwork is about twice as large as the surface-mount artwork. The surface-mount artwork includes three jumpers. On the surface-mount board, diode D1 is in a three-lead TO-23 package, with one lead unused.

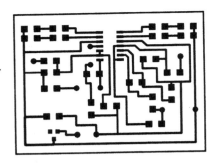

Fig. 11-18 *Actual-size pc board artwork for the logic probe, surface-mount version, top (component-side) orientation.*

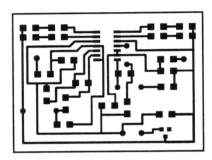

Fig. 11-19 *Actual-size pc board artwork for the logic probe, surface-mount version, mirror image.*

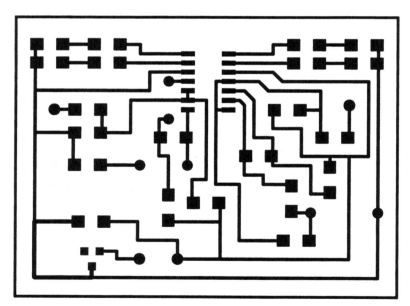

Fig. 11-20 *The pc board artwork (×2 scale) for the logic probe, surface-mount version, top (component-side) orientation.*

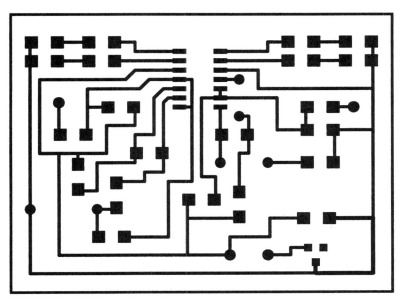

Fig. 11-21 *The pc board artwork (×2 scale) for the logic probe, surface-mount version, mirror image.*

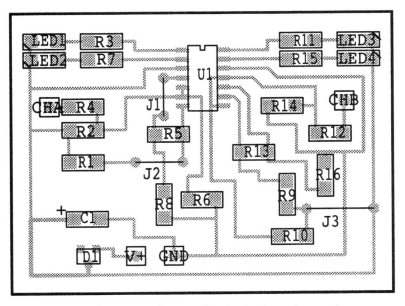

Fig. 11-22 *Parts placement diagram for the logic probe, surface-mount version.*

**Table 11-5 Parts list for
logic probe, surface-mount version.**

Resistors (⅛-W, 5% tolerance, 1206 size)

R1	120,000 Ω
R2, R6, R10	220,000 Ω
R3, R7, R11, R15	330 Ω
R4, R12	10,000 Ω
R5	200,000 Ω
R8	150,000 Ω
R9, R14	100,000 Ω
R13	470,000 Ω
R16	33,000 Ω

Capacitors (16 WVdc, 1206 size)

C1	0.1-µF tantalum

Semiconductors

D1	Small-signal diode, TO-23 package
LED1, LED3	Light-emitting diode, red
LED2, LED4	Light-emitting diode, green
U1	LM339 quad comparator, SO14 package

Miscellaneous

Test leads with alligator or mini probe tip on one end,
   banana plug on the other (4)
Banana jacks (4)

Construction: Through-hole version

Follow these simple steps:

1. To make the through-hole version of the logic probe, gather the parts listed in the parts list (Table 11-4). If you are using an enclosure with a transparent lid, or no enclosure, you can install the LEDs directly on the pc board. Otherwise you'll want to mount the LEDs on the enclosure's face and solder wires between the LEDs and their holes in the pc board.

2. Make a pc board using any of the construction methods described in this book. Drill component and mounting holes.

3. Install and solder the components in this order: IC socket, resistors, capacitor, LEDs. Be sure to observe correct polarities for LED1 through LED4 and D1.

4. Cut four lengths of hookup wire, each about 12 inches long. If possible, use four different colors of wire. Strip ¼ inch of insulation from the ends of each wire. Insert the free end of each wire into its through-hole and solder.

For front-panel-mount LEDs, solder an insulated wire to the pad for the cathode of each LED.

5. Prepare the enclosure. Drill pc board mounting holes in the floor of the enclosure. Drill holes for the four banana jacks and LED1 through LED4 in the front panel.

6. Mount the pc board in its enclosure. Thread the unconnected wires through their holes in the enclosure and solder or connect the banana jacks and LEDs to their respective wires. Wire the anodes of the LEDs to V+. Install the banana jacks and LEDs in the front panel. Insert U1 into its socket.

Construction: Surface-mount version

As with the pulser/flasher, the dual logic probe can also be built using surface-mount components. Using surface-mount components results in a tiny logic probe that you can incorporate into other circuits as a troubleshooting aid.

In building the surface-mount version, these are some differences to be aware of:

- U1 does not use a socket, but solders directly to the pc board.
- Because of the smaller components and tighter spacing, the artwork uses three jumpers (J1 through J3), which you must solder to their drilled pads using short lengths of hookup wire.
- Be sure to observe correct polarities for LED1 through LED4 and C1, as indicated on the parts placement diagram.
- Don't forget to drill through-holes for V+, GND, J1 through J3, and board-mounting holes, if used. If LED1 through LED4 will be front-panel mounted, also drill lead holes at the pads for their cathodes.
- Instead of full-size banana jacks, you might want to use miniature connnectors for the probe leads.
- Remember to use appropriate techniques for placing and soldering the tiny surface-mount devices.

Checkout and use

To check out your logic probe, plug leads into the jacks on the front panel. Connect V+ and GND to the outputs of a 5-volt supply. Connect CHA to the wiper of a 1-kilohm or greater potentiometer. Connect the other two terminals of the potentiometer to

V+ and GND on the power supply. Set a voltmeter to a scale of 5 volts or greater, and connect its leads to GND and the wiper of the potentiometer.

Vary the voltage at CHA by turning the wiper of the potentiometer. At about 1.5 volts and lower, LED2 should light green. At about 3.5 volts and higher, LED1 should light red.

Move the probe from CHA to CHB and repeat, watching LED3 and LED4 as you vary the voltage.

To use the logic probe to monitor circuit voltages, connect V+, the circuit's positive supply, and GND to the circuit's ground. Connect CHA and CHB to signals you want to monitor and watch the results on the LEDs.

Enhancements and modifications

This basic logic probe can be modified and enhanced in a number of ways including

- Add another LM339, or more than one, to increase the number of channels you can monitor.
- For an all-CMOS logic probe, eliminate R9, R13, and R16 and revise the artwork so that both halves of the circuit use the same voltage divider. To do so, connect pins 9 and 10 of U1 to pins 7 and 4 of U1. For an all-TTL logic probe, eliminate R1, R5, and R8 and revise the artwork as described previously.
- Design and add a pulse detector that will show pulses that would otherwise be too rapid to be visible on the LEDs.
- Revise the artwork so the pc board fits in a long, narrow case, for a pencil-type probe.

Appendix A: Software for drawing schematics and pc board artwork

The following are sources of software for drawing schematic diagrams and pc board artwork.

ACCEL Technologies, Inc.
6825 Flanders Drive
San Diego, CA 92121
(619) 554-1000; (800) 433-7801
FAX: (619) 554-1019
Products: Tango

Advanced Microcomputer
    Systems, Inc.
1321 NW 65th Place
Fort Lauderdale, FL 33309
(305) 975-9515; (800) 972-3733
FAX: (305) 975-9698
Products: EZ-ROUTE

Advanced Systems Design, Inc.
159 Main Dunstable Road, Suite 204
Nashua, NH 03060
(603) 882-4460
FAX: (603) 882-9223
Products: Boardmaker

BSOFT Software
444 Colton Road
Columbus, OH
(614) 491-0832
FAX: (614) 497-9971
Products: Low-cost engineering
    software

CadSoft Computer
801 S. Federal Highway
Delray Beach, FL 33483
(800) 858-8355
FAX: (407) 274-8218
Products: EAGLE

Design Computation, Inc.
Route 33
Sherman Square
Farmingdale, NJ 07727
(201) 938-6661
FAX: (201) 938-6662
Products: DC/CAD

Epoch Technology
P.O. Box 2175
Sandusky, OH 44871
Products: pc board design
    software

gav
345 W. Williams Avenue
Fallon, NV 89406
(702) 423-1653
FAX: (702) 423-1654
Products: QiCAD

Interactive CAD Systems
2352 Rambo Court
Santa Clara, CA 95054
(408) 970-0852
Products: ProCAD

Logical Systems
P.O. Box 6184
Syracuse, NY 13217-6184
(315) 478-0722
FAX: (315) 475-8460
BBS: (315) 471-3961
Products: Gerber Toolkit

Mental Automation, Inc.
5415 136th Place S.E.
Bellevue, WA 98006
(206) 641-2141
Products: SuperCAD

Number One Systems Limited
Harding Way
St. Ives
Huntingdon, Cambs.
England PE17-4WR
from USA: 011 44 480 61778
FAX from USA: 011 44 480 494042
Products: Easy-PC

Ohio Automation
7840 Angel Ridge Road
Athens, OH 45701
(614) 592-1810
Products: Easy-PC (U.S. distributor)

Old Colony Sound Lab
P.O. Box 243
Peterborough, NH 03458
(603) 924-6371
FAX: (603) 924-9467
Products: Lay01

Omation Incorporated
801 Presidential Drive
Richardson, TX 75081
(214) 231-5167; (800) 553-9119
FAX: (214) 783-9072
BBS: (214) 231-1350
Products: Schema PCB

OrCAD Systems Corporation
3175 N.W. Aloclek Drive
Hillsboro, OR 97124
(503) 690-9881
FAX: (503) 690-9891
Products: OrCAD

PADS Software, Inc.
165 Forest St.
Marlboro, MA 01752
(800) 554-7253
FAX: (508) 485-4300
Products: PADS-PCB, PADS-Logic

PCBoards
2110 14th Avenue South
Birmingham, AL 35205
(205) 933-1122
Products: PCBoards, PCRoute

Phase Three Logic, Inc.
1600 NW 167th Place
Beaverton, OR 97006
(503) 645-0313
FAX: (503) 645-0207
Products: CapFast

Protel Technology Inc.
50 Airport Parkway
San Jose, CA 95110
(408) 437-7771; (800) 544-4186
FAX: (408) 437-4913
Products: Autotrax, Easytrax

Racal-Redac
238 Littleton Road
Westford, MA 01886
(800) 356-8352
FAX: (508) 692-4725
Products: MAXI/PC

Sector Systems Company
416 Ocean Avenue
Marblehead, MA 01945
(617) 639-2625
Products: Catalog of engineering
    public domain software

Softronix
14 Greenbrook Lane
Kingston, NY 12401
BBS: (914) 336-2744
Products: EditSCH, EditPCBI

Suncoast Technologies
P.O. Box 5835
Spring Hill, FL 34606
(904) 596-7599
Products: Inexpensive pc board
    software

Synergetics
P.O. Box 809
Thatcher, AZ 85552
(602) 428-4073
Products: pc board design using
    Postscript

Tsien (UK) Limited
Cambridge Research Laboratories
181A Huntingdon Road
Cambridge CB30DJ
United Kingdom
Products: Boardmaker

Vamp
6733 Selma Avenue
Los Angeles, CA 90028
(213) 466-5533
Products: McCAD for Macintosh

Visionics Corporation
3032 Bunker Hill Lane
Suite 201
Santa Clara, CA 95054
(408) 492-1440; (800) 553-1177
FAX: (408) 492-1380
Products: EE Designer

Wintek Corporation
1801 South Street
Lafayette, IN 47904-2993
(317) 742-8428; (800) 742-6809
FAX: (317) 448-4823
Products: Hiwire, Smartwork

General-purpose CAD software

These are some sources for moderate-cost, general-purpose CAD software that you can use to draw schematics and pc board artwork

American Small Business
    Computers, Inc.
327 South Mill
Pryor, OK 74361
(918) 825-4844
FAX: (918) 825-6359
Products: DesignCAD 2D

Autodesk Retail Products
11911 North Creek Parkway South
Bothell, WA 98011
(206) 487-2233
FAX: (206) 485-0021
Products: Generic CADD

Foresight Resources Corp.
10725 Ambassador Drive
Kansas City, MO 64153
(816) 891-1040
Products: Drafix Windows CAD

International Microcomputer
    Software, Inc.
1938 Fourth Street
San Rafael, CA 94901
(415) 454-7101; (800) 833-4674
Products: TurboCAD

Appendix B: Where to buy supplies and materials

Where can you buy the supplies, tools, and equipment you need for pc-board fabrication? The following list includes a variety of sources.

Product manufacturers are good sources for detailed information and advice about their products. Some manufacturers also sell directly by mail, while others will refer you to distributors or sales representatives.

Also listed are mail-order vendors who offer a range of products related to pc boards, from a variety of manufacturers.

Advance Process Supply Company
400 N. Noble Street
Chicago, IL 60622-6383
(312) 829-1400; (800) 678-1900
FAX: (312) 942-8590
Screen printing supplies

All Electronics Corp.
P.O. Box 567
Van Nuys, CA 91408-0567
(800) 826-5432
FAX: (818) 781-2653
pc board materials, components

Bishop Graphics
5210 Lewis Road
P.O. Box 6012
Agoura Hills, CA 91376-6012
(818) 991-2600
FAX: (818) 889-3744
pc board drafting aids

Black Lightning
Riddle Pond Road
West Topham, VT 05086

(802) 439-6462; (800) 252-2599
FAX: (802) 439-6463
Toner transfer materials, ultra-black
    toner cartridges

Capital Advanced Technologies
309-A Village Drive
Carol Stream, IL 60188
(708) 690-1696
FAX: (708) 690-2498
Surface-mount prototyping supplies

Chemtronics Inc.
8125 Cobb Centre Drive
Kennesaw, GA 30144
(404) 424-4888; (800) 645-5244
FAX: (404) 424-4267
Chemicals for electronic production
    and field service

Contact East, Inc.
335 Willow Street
North Andover, MA 01845-5995
(508) 682-2000
FAX: (508) 688-7829

Contact East, Inc. (*continued*)
Products for electronics test, repair,
and assembly

Cooper Tools
P.O. Box 951384
Dallas, TX 75395
(919) 362-1709
Weller soldering tools

Datak Corporation
55 Freeport Boulevard, Unit #23
Sparks, NV 89431
(702) 359-7474
FAX: (702) 359-7494
Dry transfer patterns, pos-neg film,
pc board supplies

DC Electronics
P.O. Box 3203
Scottsdale, AZ 85271-3203
(800) 423-0070
FAX: (602) 994-1707
PC boards and supplies, enclosures

Digi-Key Corporation
701 Brooks Avenue South
P.O. Box 677
Thief River Falls, MN 56701-0677
(800) 344-4539
Wide variety of electronic components
and related items

Dremel
P.O. Box 1468
Racine, WI 53401-9989
Manufacturer of high-speed rotary
tools

DynaArt Designs
3535 Stillmeadow Lane
Lancaster, CA 93536
(805) 943-4746
Iron-on toner transfer paper

ESP Solder Plus
14 Blackstone Valley Place
Lincoln, RI 02865
(401) 333-3800
FAX: (401) 333-4954
Solder pastes, dispensers, and
related products

Excellon Automation Co.
24751 Crenshaw Boulevard
Torrance, CA 90505
(213) 534-6300
FAX: (213) 530-8620
Manufacturer of automated
pc board drilling systems

GC/Thorson (GC Electronics)
1801 Morgan Street
Rockford, IL 61102
(815) 968-9661
FAX: (815) 968-9731
pc boards and related materials and
equipment

General Consulting
12856 Inglewood Avenue
Hawthorne, CA 90250
(213) 679-1405
Fast-Proto direct-plot materials

Gerber Scientific Instrument
Company
83 Gerber Road West
South Windsor, CT 06074
(203) 644-1551
Photoplotter manufacturer

Hosfelt Electronics
2700 Sunset Boulevard
Steubenville, OH 43952-1158
(800) 524-6464
FAX: (614) 264-5414
Surplus distributor, resharpened
drill bits

International Eyelets, Inc.
5674 El Camino Real, Suite L
Carlsbad, CA 92008
(800) 333-9353
Eyelets for double-sided boards

Jameco
1355 Shoreway Road
Belmont, CA 94002
(800) 831-4242
FAX: (800) 237-6948
Electronic components and tools

JDR Microdevices
2233 Samaritan Drive
San Jose, CA 95124
(800) 538-5000
FAX: (800) 538-5005
Electronic components and tools

Jensen Tools, Inc.
7815 S. 46th Street
Phoenix, AZ 85044-5399
(602) 968-6231
FAX: (602) 438-1690; (800) 366-9662
Products for electronics testing, repair,
    and assembly

Kepro Circuit Systems
630 Axminister Drive
Fenton, MO 63026-2992
(314) 343-1630; (800) 325-3878
FAX: (314) 343-0668
Wide variety of pc boards and related
    materials and equipment

Litton/Kester Solder
515 East Touhy
Des Plaines, IL 60018
(708) 297-1600
Solder and chemicals

Meadowlake Corporation
P.O. Box 497
Northport, NY 11768
TEC-200 iron-on transfer film

Mouser Electronics
2401 Highway 287 North
Mansfield, TX 76063-4827
(817) 483-4422; (800) 346-6873
FAX: (817) 483-0931
Electronic parts, tools, and supplies

Multicore Solders
1751 Jay Ell Drive
Richardson, TX 75081
(214) 238-1224
FAX: (214) 437-0288
Solders and fluxes

Newark Electronics (branches
    nationwide)
4801 N. Ravenswood Avenue
Chicago, IL 60640-4496
(312) 784-5100
pc boards, related tools and
    materials, electronic
    components

OK Industries
4 Executive Plaza
Yonkers, NY 10701
(914) 969-6800
FAX: (914) 969-6650
Tools and equipment for soldering
    and pc board repair

PACE, Inc.
9893 Brewers Court
Laurel, MD 20723
(301) 490-9860
FAX: (301) 498-3252
pc board repair products

Planned Products
303 Potrero Street, Suite 53
Santa Cruz, CA 95060
(408) 459-8088
FAX: (408) 459-0426
Manufacturer of Circuit Works
    conductive pen, conductive
    epoxy, and overcoat pen

Porter's Camera Store, Inc.
Box 628
Cedar Falls, IA 50613
(319) 268-0104; (800) 553-2001
FAX: (800) 779-5254
Wide variety of photographic
    equipment and supplies

Project Pro
1710 Enterprise Parkway
Twinsburg, OH 44087
(800) 800-3321
FAX: (216) 425-1228
Project enclosures

Prototype Systems
12930 Saratoga Avenue
Suite D1
Saratoga, CA 95070
(408) 996-7401
FAX: (408) 996-7871
Equipment and materials for pc board
    fabrication, including positive-
    acting, aqueous-process
    presensitized pc boards

Radio Shack (stores nationwide)
Division of Tandy Corporation
Fort Worth, TX 76102
Basic pc board materials

Small Parts Inc.
13980 N.W. 58th Court
P.O. Box 4650
Miami Lakes, FL 33014-0650
(305) 557-8222
FAX: (800) 423-9009
Drill bits, hardware

Solder World
9555 Owensmouth Avenue, #14
Chatsworth, CA 91311
(818) 998-0627
FAX: (818) 709-2605
pc boards, supplies

Specialized Products Company
3131 Premier Drive
Irving, TX 75063
(214) 550-1923; (800) 866-5353
FAX: (800) 234-8286
Mail-order source fof products for
    testing, repair, and assembly

Techniks Inc.
P.O. Box 463
Ringoes, NJ 08551-0463
(908) 788-8249
Press-n-Peel iron-on transfer film

Techni-Tool
5 Apollo Road
P.O. Box 368
Plymouth Meeting, PA 19462
(215) 941-2400; (800) 832-4866
FAX: (215) 828-5623
Electronic tools and supplies

TECH SPRAY Inc.
P.O. Box 949
Amarillo, TX 79105-0949
(806) 372-8523
Chemicals for electronic production
    and field service

Think & Tinker Ltd.
Box 408
Monument, CO 80132-0408
(719) 488-9640
No-etch fabrication techniques

3M
3M Center Building
St. Paul, MN 55144
Scotchflex breadboard kits,
    conductive adhesive transfer
    tape

Ulano
255 Butler Street
Brooklyn, NY 11217
(718) 622-5200
FAX: (718) 802-1119
Screen-printing supplies

Ungar
5620 Knott Avenue
Buena Park, California 90621
(714) 994-2510
FAX: (714) 523-7790
Soldering tools

Appendix C: Information sources

For more information about pc board design and fabrication, try the following books, magazines, and organizations.

Books

If you want to learn more about designing or fabricating pc boards, the books listed here are a good place to start. Many local bookstores will special-order titles on request at no additional charge. Some publishers sell directly by mail. Also check your local public, university, or technical college libraries for copies of books you're interested in. If your library doesn't have the titles you want, see if it can obtain copies for you on interlibrary loan.

Analog Printed Circuit Design and Drafting
Darryl Lindsey
1985
Bishop Graphics
Learn-by-doing text with examples

Beginner's Guide to Reading Schematics
Robert J. Traister and Anna L. Lisk
1986
TAB Books

Design Guidelines for Surface Mount Technology
Vern Solberg
1992
TAB Books

Design Guidelines for Surface Mount Technology
John E. Traister
1990
Academic Press

Digital Printed Circuit Design and Drafting
Darryl Lindsey
1986
Bishop Graphics
Learn-by-doing text with examples

Electronic Drafting and Printed Circuit Board Design
James M. Kirkpatrick
1985
Delmar Publishers
Drafting basics

*Electronic Packaging and
    Interconnection Handbook*
Charles A. Harper
1991
McGraw-Hill

*Handbook of Surface Mount
    Technology*
Stephen W. Hinch
1988
John Wiley & Sons

*How to Draw Schematics and Design
    Circuit Boards with Your IBM PC*
Steve Sokolowski
1988
TAB Books
Includes BASIC programs for drawing
    schematics and pc board artwork

*How to Read Electronic Circuit
    Diagrams,* 2nd Edition
Robert M. Brown, Paul Lawrence, and
    James A. Whitson
1987
TAB Books

How to Read Schematics, 4th Edition
Donald E. Harrington
1986
SAMS

*KODAK Index to Photographic
    Information* (publication L-1)
Eastman Kodak Company, Dept. 412-L
343 State Street
Rochester, NY 14650-0532
Lists over 400 Kodak publications on
    photographic topics

*North American Directory of Contract
    Manufacturers in Electronics*
Updated yearly
Miller Freeman Publications
Lists pc board designers,
    manufacturers, assemblers

Printed Circuit Board Basics, 2nd
    Edition
Michael Flatt
1991
Miller Freeman Books
Introduction to the pc board industry

*Printed Circuit Board Design with
    Microcomputers*
T.J. Byers
1991
TAB Books
Includes reviews of selected software

*Printed Circuit Board Precision
    Artwork Generation and
    Manufacturing Methods*
Preben Lund
1986
Bishop Graphics

*Printed Circuit Engineering:
    Optimizing for Manufacturability*
Raymond H. Clark
1988
Van Nostrand Reinhold

Printed Circuits Design
Gerald L. Ginsberg
1991
McGraw-Hill
How to design pc board artwork

Printed Circuits Handbook, 3rd
    Edition
Clyde F. Coombs, Jr., Editor-in-chief
1988
McGraw-Hill
Exhaustive detail on pc board
    manufacturing, with the emphasis
    on commercial fabrication

*SMT Guidelines for Printed Circuit
    Board Design*
James Blankenhorn
1988
Bishop Graphics

*Soldering Handbook for Printed
    Circuits and Surface Mounting*
H. Manko
1986
Van Nostrand Reinhold

*Solder Paste Technology: Principles
    and Applications*
Colin C. Johnson and Joseph Kevra,
    Ph.D.
1992
TAB Books

Surface Mount Technology
Carmen Capillo
1991
McGraw-Hill

Surface-Mount Technology for PC-
*   Board Design*
James K. Hollowman, Jr.
1992
SAMS

Surface Mount Technology Handbook
Vern Solberg
1990
TAB Books

Surface Mount Technology:
*   Principles and Practices*
Ray P. Prasad
1989
Van Nostrand Reinhold
Technical and applications
    information for working with
    surface-mount devices

Surface Mount Technology: The
*   Handbook of Materials and*
*   Methods*
Bernard S. Matisoff
1989
TAB Books

Magazines

Magazines are good places to learn about the latest developments
in processes and products relating to pc board design and fabrication. Some of the magazines listed below are dedicated to covering topics relating to pc board design and fabrication. Others
have a broader focus but include occasional articles and frequent
advertisements for products relating to pc boards.

Magazines described as "free to qualified subscribers" are
financed by their advertisers. Potential subscribers are asked to
fill out a qualification card describing their involvement with the
types of products the magazine covers. Respondents who indicate that they buy or influence the buying of sufficient quantities
of the products are granted a free subscription. Paid subscrip-
tions are also available, though usually at high cost. To find out
more about a publication, call or write for a sample issue.

Circuits Assembly
Miller Freeman Inc.
600 Harrison Street
San Francisco, CA 94017
Free to qualified subscribers; focuses
    on pc board assembly

The Computer Applications Journal
4 Park Street
Vernon, CT 06066
(203) 875-2751
BBS: (203) 871-1988
Computer topics including pc board
    design and fabrication

ComputerCraft
76 North Broadway
Hicksville, NY 11801
(516) 681-2922
FAX: (516) 681-2926
Computer hardware and software
    including pc board design and
    fabrication

EDN
275 Washington Street
Newton, MA 02158
Free to qualified subscribers;
    electronic technology for
    engineers

Electronic Design
1100 Superior Avenue
Cleveland, OH 44114-2543
(216) 696-7000
Free to qualified subscribers; for
    engineers and engineering managers

Electronic Engineering Times
600 Community Drive
Manhassat, NY 11030
Free to qualified subscribers; news
    about electronic products

Electronic Servicing and Technology
P.O. Box 12487
Overland Park, KS 66212
Servicing and repair

Electronics Handbook
P.O. Box 5148
North Branch, NJ 08876
Hobby electronics

Electronics Now
500-B Bi-County Boulevard
Farmingdale, NY 11735
(516) 293-3000
Hobby electronics

Midnight Engineering
1700 Washington Avenue
Rocky Ford, CO 81067-2246

(719) 254-4558
Entrepreneurial electronics

Nuts & Volts
430 Princeland Court
Corona, CA 91719
(714) 371-8497
FAX: (714) 371-3052
Electronics shopper magazine

Popular Electronics
500-B Bi-County Boulevard
Farmingdale, NY 11735
(516) 293-3000
Hobby electronics

Printed Circuit Design
Miller Freeman Inc.
600 Harrison Street
San Francisco, CA 94017
Free to qualified subscribers;
    focuses on pc board design

Surface Mount Technology
17730 West Peterson Road, Box 159
Libertyville, IL 60048-0159
(708) 362-8711
FAX: (708) 362-3484
Free to qualified subscribers;
    surface-mount components and
    pc boards

Organizations

The following organizations publish standards and other guidelines relating to pc board design and fabrication

American National Standards Institute
    (ANSI)
1430 Broadway
New York, NY 10018
(212) 354-3300
Publishes standards relating to
    electronic components

Electronics Industries Association (EIA)
2001 Eye Street NW
Washington, DC 20006
(202) 457-4930
Publishes JEDEC (Joint Electron Device

Engineering Council) standards
    for mechanical outlines of
    electronic components

Institute for Interconnecting and
    Packaging of Electronic Circuits
    (IPC)
7380 North Lincoln Avenue
Lincolnwood, IL 60646
(708) 677-2850
FAX: (708) 677-9570
Publications and videos on pc board
    design and fabrication

Index